PRAISE FOR TWILIGHT IN THE 'NAM

"It's been 48 years since I served in the 196th Light Infantry Brigade as a 20/21 year-old grenadier in Lieutenant Sherwood's platoon. *Twilight in the 'Nam* brought back a shovel full of memories and emotions that have been boxed and stored away for a very long time.

Sherwood's descriptions of the variety of personalities in our unit and how they dealt with various situations are quite accurate. Then there were the vivid reminders of the relentless monsoon rains, the 'skeeters, heat and humidity, and the cumbersome rucksacks that were filled with ammo, grenades, claymore mines, and C-4. His descriptions of the many weeks in the field, oscillating between adrenalin spikes and misery and monotony, between laughter and frustration, were evocative.

This book created a craving to relive those days that we, at the time, so desperately wanted to avoid.

Finally, *Twilight in the 'Nam* documents a young lieutenant's views and responsibilities in the jungles of Vietnam."

— Ruben Bugge, Taos New Mexico
SP4, Co C, 1st Bn., 46th Inf, 196th Light Infantry Brigade

"The author takes you into the middle of an Infantry Platoon in the Vietnam War. You will begin to see through his eyes the men he served with and those who commanded him as well. You will see the looks on the men's faces he served with as they search for the enemy with anticipation and then boredom as nothing is found for days on end, and then how quickly things change as signs of the fleeting enemy appear without warning.

You will feel the cold rain running down your neck, the weight of your rucksack on your back, and the sharp vines wrapping around your legs as you try and make your way quietly through dense jungles and up and down steep slopes in search of the elusive enemy. And you will gain understanding about what it is like to sense the movement of some unknown, unseen creature as it crawls up your arm as you wait in complete stillness in a night ambush position on the darkest of nights.

Most importantly, you will begin to understand the weight of responsibility that this officer, like those in the past, present, and future, carries on his shoulders each and every day.

This is a book that every Infantryman can relate to, regardless of which war he served in. And it is a book that, if you have not served in the Infantry, you will begin to gain a better appreciation for those of us who have."

—LTC (Ret) Ed DeVos
Advisor, 21st ARVN Division, 1972

TWILIGHT IN THE 'NAM

TWILIGHT IN THE 'NAM

WITH THE 196TH LIGHT INFANTRY BRIGADE

BYRNE N. SHERWOOD

Deeds Publishing | Athens

Copyright © 2020 — Byrne N. Sherwood

ALL RIGHTS RESERVED—No part of this book may be reproduced in any form or by any electronic or mechanical means, including information storage and retrieval systems, without permission in writing from the authors, except by a reviewer who may quote brief passages in a review.

Published by Deeds Publishing in Athens, GA
www.deedspublishing.com

Printed in The United States of America

Cover design by Mark Babcock.

ISBN 978-1-947309-87-6

Books are available in quantity for promotional or premium use. For information, email info@deedspublishing.com.

First Edition, 2020

10 9 8 7 6 5 4 3 2 1

To the soldiers of:

1st Platoon, Co C, 1st Battalion, 46th Infantry,
196th Light Infantry Brigade
and

The Boys of El Cerrity
When I was young and Walter Mitty
I dreamt of deeds of valor
For one of El Cerrito's ne'er do wells,
a little bit of honor
Instead I am counted among my peers
My heroes, those who served
To have participated in their sacrifice
Is more than I deserve

Rudy Carriaga WIA 1st Cav Div
Denny Derrer 1st Inf Div
Andy Garnica KIA 25th Inf Div
Jim Graham USMC
Pat Hilliard 1st Inf DIv
Tyrone McDade WIA x2 9th Inf Div
Mike Mendoza WIA 101st Abn Div
Craig Mosher
Roger Oliver WIA
Don Righter
Ralph Stacy KIA 173d Abn Bde
John Stone WIA x3 1st Cav Div
James Worth 101st Abn Div
Tony Zanotelli WIA 199th LIB

For my Grandson Noah

I hope that someday you will, in the company of others, be pushed to your physical, mental and emotional limits in pursuit of a worthy cause that is greater than yourself; for it is in the pouring out of ourselves that we live most fully. I hope that you will experience the comradeship that accompanies such an endeavor. But not in war.

CONTENTS

Foreword	xiii
Acknowledgements	xv

1. Good Morning Vietnam!	1
2. REPL DEPL	5
3. Welcome To Professional Gas	9
4 Early Days	13
5. Preparation	21
6. 8 January 1972	27
7. Booby Trap	33
8. Mountains	39
9. Comings and Goings	45
10. Hills, Hills and More Hills	49
11. Re-supply, An Unexpected Gift and Gas	55
12. A Bath and a Home Cooked Meal	61
13. Monsoon R & R	65
14. Spider-Man	83
15. Mail	87
16. MA	89
17. Eagle Flights	95
18. Dominoes	103
19. First Contact?	107
20. Ridgeline	111
21. Settling In	117
22. New Battalion Commander	121
23. Boredom and Drugs	125
24. More Ridgeline and a Dining-in	129
25. I Contemplate Murder	133

26. War on Drugs	137
27. Farewell to the Ridge	141
28. Charlie Ridge	145
29. Recon Ridge	155
30. Pushing the Boundaries	165
31. Some Like it Hot, Some Like it Cold	175
32. What Are You Going to Do Now, Lieutenant?	179
33. NVA Easter Offensive	183
34. Jungle Sores & Friendly Fire	185
35. Uncertain Future	189
36. Rumors, Replacements and a Bath	195
37. Two Weeks Easy Duty?	203
38. Fire Base	207
39. Mystery Pajamas	213
40. Settling In	219
41. Artillery Raid	227
42. Inspection and Ho Chi Minh's Birthday	233
43. Life on the Firebase	237
44. Visitors	243
45. Eureka	247
46. Good Night And Good Luck	259
Appendix A	267
Appendix B	277
Appendix C	281
About the Author	285

FOREWORD

Why another Vietnam memoir? Surely enough have been written. My stated reason to myself and others was that I wanted to write it for my grandson. The more I have thought about it, I would have to say that while that is not untrue, it is only partially true.

Most histories of the Vietnam War devote a page or two to the last year of U.S. ground force involvement in Vietnam. Always present in those one or two pages are references to rampant drug use, poor discipline, low morale, racial tension, fragging, and a host of other negatives, basically describing a disintegrated army. While I can't deny any of that, I can say that none of that existed in my platoon or in any other platoon of which I was aware.

Furthermore, President Nixon informed the American public that, at that stage of withdrawal from Vietnam, the only mission that soldiers in Vietnam had was to guard the two airbases at Ton Son Nhut and Da Nang. While not technically untrue, the image it created of soldiers walking guard on a fence line around an airfield was completely false. In reality, we were humping the mountains twenty kilometers to the west in an effort to prevent the NVA from launching rocket attacks on those airfields.

So, I guess my deepest reason for writing a memoir about that last, unglamorous year of U.S. ground force involvement in Vietnam is to pay tribute to those young American G.I.'s who, with no hope of victory, were still willing to put themselves in harm's way in the service of their country. Why were they? I don't really know. Loyalty to comrades, self-respect, pride, all the above? But whatever the reason, what is true is that they will always have my undying respect and admiration. It was my greatest honor to serve with them.

ACKNOWLEDGEMENTS

If not for the writing group to which I belong, my book would have read like a military after-action report since this was the style of writing I had employed throughout my professional life. Overwhelmingly without military experience, they wanted more than facts: they wanted emotions, sights, sounds, smells, tastes, and touch. I am indebted to them for breaking me out of a sterile, fact driven style of writing. I must name them, for each inspired me in their own way: John and Cherry Benzie, Duane Berger, Christine Chu, Larry Crevin, Werner Gottlieb, Judy Hammon, Karl Klausner, Catalina Sanders, Heidi Schmidt, Lisa Stepp, and Shelley Waldenberg.

I have to thank my parents for saving my letters, which I discovered among their effects after they died. Without them, my memories would be a jumble of individual recollections floating independently in my mind without sequence or timeline.

I also want to thank John Beasley, Ruben Bugge, Peter Gadzinski, and Roberto Granado, comrades who provided clarification and encouragement.

Finally, I want to thank Bob Babcock, my publisher and Vietnam veteran, for his encouragement and peerless editing.

TWILIGHT IN THE 'NAM

1. GOOD MORNING VIETNAM!

Finally, here it was. I was on my way to the experience I craved above all others. I was seated on a Flying Tiger chartered airliner headed for Vietnam, something I had been waiting for since 1965 when U.S. ground forces were first deployed there and consumed so many of my friends and acquaintances from high school. Men fought their country's wars, as had my father and other male ancestors before him. And now it was my turn.

Mixed with my excitement was a somberness brought on by the reality of what I had gotten myself into. By 1971 there were no longer any illusions that this was a crusade to make the world safe for freedom and democracy. It was a brutal war and one in which the U.S. was not going to prevail. In addition, I was leaving behind a pregnant wife who I had lied to about the fact that I had volunteered for this, not once but several times.

After the usual preliminaries, the plane took off from Travis Air Force Base and headed west. Very soon, we were back over the San Francisco Bay Area, where the plane banked right and headed north. Out of the left window I looked down at the Golden Gate Bridge and I told myself to take a good, long look because it might be the last look I would have of the place where I grew up and which held so many memories for me. During that long flight, I took the opportunity to contemplate, and ultimately accept, my own death. That philosophical exercise gave me a different outlook on life, an outlook that has remained with me ever since and is captured by altering slightly the G.I. saying that was often engraved on Zippo lighters, "For those who have almost lost it, life has a flavor that the protected will never know."

Back in touch with the reality of my immediate surroundings, I was struck by how indifferently we were treated by the flight crew. I guess I expected some sympathy from the stewardesses but, instead, we were treated no differently than if we were civilians on our way to some routine business meeting. Several hours later we landed in Anchorage, Alaska. We were let off the plane and into the terminal where I made my way to the bar to kill time until re-boarding. Here I found the kind of sympathetic reception that I hoped for from the flight crew. The locals in the bar gave us sympathy and good wishes, made tangible by buying us beers.

When we re-boarded the plane, I discovered that we had a new crew. If the stewardesses on the leg from Travis to Anchorage had been indifferent, this lot was downright callous. I felt like I had gone from being the equivalent of a civilian on his way to a business meeting to being like a side of beef being hauled to market. For one thing, this was still in the day when air travel was relatively young and being a stewardess was considered to be a glamorous occupation. In my limited experience, most stewardesses were young and attractive, but not these. They were neither young nor attractive. I learned later that the run into Vietnam was the "money run", paying more than normal routes and that, accordingly, it was reserved for those with the most seniority. Looking back in charity, I have to recognize that, whereas this was my first trip to Vietnam, it was a trip these ladies had made hundreds of times before. Any sympathy they may have felt for those on their way to Vietnam early on must by now have been worn away by long repetition.

The next stop on our journey was Yokota, Japan, where we again de-planed and hung around the terminal. I only have two recollections of this. The first was a feeling of excitement at being in a foreign country. "Wow, here I am in Japan. I was supposed to come live here when I was four and now, after that aborted mission, here I finally am." The second and almost immediate thought was that my immediate surroundings looked no different than any other airport that I had ever been in. So much for Japan.

Back on the plane, next stop Vietnam—but not for a while. Japan may be in Asia but it's still a long way from there to Vietnam. Finally, the pilot announced that we were approaching our destination, which was Ton Son Nhut Air Base, outside Saigon. I immediately became fully alert and started looking intently out the window. We were still in the predawn hours of complete darkness, which allowed me to see the lit-up perimeters of fire-bases amidst the blackness of what I took to be the surrounding jungle.

Just before the break of dawn, the plane landed and taxied to the point where we were to disembark. As we stepped off the plane, I was assaulted by two things simultaneously: the intense, suffocating, hot, humid air and the incredible stench. I think that everyone who ever landed there would describe the same two sensations. It's hard to describe a smell but I would say that it was the smell of putrefaction—rotten vegetables, rotten fish, rotten everything. Perhaps the only comparison that I could make would be to compare it to walking down Bourbon Street in New Orleans on a hot, summer morning, with the previous night's swill and oyster shells simmering together in the hot sun. Magnify that by one hundred per cent and you may have something close.

In the grayness of the breaking dawn, we reported to a window where we surrendered our U.S. money and received military scrip in return. Next I went to use the bathroom to take a leak and tidy myself up a bit. In the men's bathroom was an old mama-san sweeping the floor with a little broom that looked like a palm frond. I waited a moment for her to leave so that I could do my business in private but it quickly became apparent that she had no interest in my business or my privacy. I thought, "Well, if it's OK with her, it's OK with me." It was quickly dawning on me that I was indeed in a place unlike any that I had ever experienced before.

We then passed through a gate onto a wide road and lined up to get on a bus that would take us to the replacement depot. The sun was just peeping up in the sky now and, as I waited my turn to get on the bus, I heard a high-pitched buzzing sound like a swarm of huge insects. Looking down the road to my left, I saw heading toward us what looked like a tsunami of the ubiquitous three-wheeled motorized cyclos. A new day was beginning in 'The Nam."

2. REPL DEPL

We lined up at the gate of Ton San Nhut to board a bus which would take us to the replacement depot. The bus was a standard Army bus, or I should say, a standard school bus painted olive green, with the added feature of chain link screens over the windows to prevent a grenade or a satchel charge being thrown in. I don't remember much about the ride to the 90th Replacement Detachment in Long Binh, except for the fascination of this different world, this sea of strange humanity with all its equally strange sights and smells. It looked just like pictures I had seen before, but to experience it with all the senses activated was something else altogether.

The replacement detachment, as the name suggests, was a place to hold new arrivals until they could be matched against a request for replacements somewhere in Vietnam. The officer's hut to which I was shown was right next to the perimeter fence, an easy hand grenade toss from the road. In my khakis and without a weapon, I felt very naked and vulnerable. I selected an unoccupied bunk, threw my kit bag down and made my way to the building to which we were told to report. We received several briefings, none of which I remember except that we were instructed to check the Up-country Shipping Roster every day until our name appeared with our assignment.

Next came the issue of jungle fatigues and jungle boots. Back at the hut, I changed out of my khakis and into my brand-new fatigues and boots, which screamed to the world that I was a "newbie". Like a little boy with new shoes, I wished I could find a mud puddle where I could break my boots in and not have them look so embarrassingly new. Now "on my own time", as the Army saying goes, I strolled down to the tiny Post Exchange and looked over the merchandise, which took all of ten minutes. There was nothing there that I needed or wanted. It was mid-morning now and there was nothing else to do.

What to do now? It was hotter than hell and, if I had to kill time, I at least wanted to do it in some comfort. I realized that the Officer's Club was probably air-conditioned and, even if it wasn't, they would have cold beer to help counter the heat. Up I went to the O Club, which was near the officer's hut, only to find that it didn't open until 1100. I was doomed to swelter for another hour until opening time.

The club was large and airy, with a central bar and tables and chairs scattered around. Working there were a handful of Vietnamese girls about my age, dressed in traditional loose trousers and loose fitting blouse. I was quickly charmed by their simplicity and friendliness. While hanging around there I asked them to teach me Vietnamese, which they seemed delighted to do. They taught me some everyday phrases, some of which I still remember. I went over there with a very negative attitude toward the Vietnamese people. In my youthful macho arrogance, it seemed like the way to be. The undoing of that began on the first day in country with the kindness of these girls. With nothing else to do, I hung around until closing time, which was probably 9:00 or 10:00 p.m.

The next morning, I woke up, performed my morning ablutions, and went to chow. Following chow, I went to the PX and looked at the same stuff I had looked at the day before. Next I looked at the Up-country Shipping Roster and took note of the fact that my name was not on it. Nothing to do now but wait until 1100 when the Officer's Club opened and commence drinking beer in a feeble attempt to beat the heat. Some time that night I would stumble back to the officer's hut and crash, only to repeat the same sequence the following day. I discovered that beer and heat don't mix well so I cut back on the number off beers that I consumed during the day. I felt as though I had been sentenced to death by boredom and that this was to be my routine for eternity. How long could this go on?

During the trip over, I met a helicopter pilot who was a Cajun from Louisiana and we began hanging around together. He was on his second or third tour in Vietnam and was a bit of a wild man. One night in the club after drinking all day, we unsuccessfully tried to pick a fight with some West Pointers. Failing at that, we decided to fight each other—open handed. We were moving all over the club and were having a grand time of it, but the bar girls were very upset because they thought we were having or were going to have a real fight. I remember being struck by their gentleness and their genuine desire that there not be any violence. Finally, we gave in to their pleading and settled down. That night after the club closed, I fell asleep on the sand bags outside our hut while my Cajun friend was telling his life story to someone we didn't even know.

The next day I woke up with a mighty hangover and a cold. Having nothing else to do I appeared at the Officer's Club at 1100. The same girls were working and they chided me playfully.

"You were vewy dwunk las night. You ack like cwazy man, like dinky dau."

"Yeah, well I'm not feeling so good today. I feel number ten."

"You sit."

Dutifully I plopped down in a nearby chair. One of the girls stood behind me and massaged my temples with her fingertips. Then she administered the most amazing massage therapy: holding her hands together in the prayer position but with fingers spread apart, she began hitting me with the blade of her hands all over my head, neck and shoulders. At each chop her fingers would come together and make a popping noise. It was unbelievably soothing.

"How you feewl?"

"Much, much better. Thank you."

At the time and even in my memory, it seemed like I was at the 90th Replacement Detachment for a long time. I felt like I was going to languish there for my whole tour of duty. In fact, on my third day there my name appeared on the Up-country Shipping Roster, assigned to the 196th Light Infantry Brigade in I Corps zone, with headquarters just outside of Da Nang. I had hoped to be assigned to the 1st Cavalry Division or the 101st Airborne Division but it was clear that this was going to be my assignment and I was anxious to get going. So, on the 20th of December, I shouldered my kit bag for the movement north.

3. WELCOME TO PROFESSIONAL GAS

A SUBSIDIARY OF CHARGER PETROLEUM

On Wednesday, the 22nd of December, we headed north to Da Nang by cargo plane via Ben Hoa Air Base and the base at Cam Ranh Bay. Shortly after arriving at the Cam Ranh Bay terminal, I was treated to the spectacle of several soldiers being released from a drug rehabilitation unit. They didn't look very rehabilitated to me. They came spilling out of a room with unkempt uniforms, long hair, and spaced-out looks on their faces. One in particular, with several strands of colored hippie beads around his neck, broke into what appeared to be his version of an Apache war dance while singing "rifle rounds, mortar shells, cannons all the way…" to the tune of 'Jingle Bells'. I was pretty un-nerved. It seemed to confirm all the negative press about rampant drug use by soldiers in Vietnam. I thought, "Oh God, is this what I am going to be facing when I take over a platoon?"

For its shock value, it reminded me of an incident when I was at Ft. Lewis for my basic training in the summer of 1969. We had just arrived and were at the Central Issue Facility to draw our uniforms and equipment. As we stood there at the issue counter, up rolled a bus full of soldiers just returned from Vietnam. As they filed off the bus, I was struck by their faded-out jungle fatigues and their lean and wild appearance. I especially remember one individual, a tall African

American wearing the shoulder patch of the 101st Airborne Division. The Combat Infantryman Badge pinned to the front of his cap indicated that he had been in the thick of the fighting. As he disdainfully looked us over in our brand-new fatigue uniforms, he yelled out, "You mothafuckas are going to die in Vietnam."

One of the peculiarities of being in the Army is that, in spite of its vast size, one is always bumping into an acquaintance. While waiting at Cam Ranh for my flight north and passing the time by watching people come and go, I spied a familiar face. Not just a familiar face, but the face of a friend.

"Beas!" I called out.

"Buzz!" was the equally enthusiastic response.

It was John Beasley, one of my best Army buddies. We had been in the same platoon in Ranger School and in the same company (B Co., 2d Bn., 508th Abn Inf) in the 82d Airborne Division at Fort Bragg, North Carolina prior to receiving orders for Vietnam. He had gone to Vietnam a few months ahead of me and was now in the process of transferring to another unit.

"So, what are you up to?" I asked.

"Oh man, it's been crazy. My battalion stood down and I am being transferred to Hoi An."

"No shit. Where is Hoi An?"

"About thirty klicks south of Da Nang. I'll be doing liaison work with the ARVN."

"Well that sucks. What will you be doing?"

"It's a REMF job. I'll be involved in the Pacification Program in the villages around there. I'll be visiting different sites to see how they are doing. I took a jeep when I left my last unit so it's pretty easy to get around having my own wheels."

"You stole a jeep!?"

"Well, I didn't really steal it, I just reallocated government property."

We both laughed at this common euphemism for theft of Army equipment.

We spent most of the day together and it really brightened me up to stumble into this chance encounter. John gave me his in-country postal address and we agreed to stay in touch.

I finally got back on a plane for Da Nang Air Base, arriving there at night. It was monsoon season in that part of the country and it was surprisingly chilly. I could tell that we were on the outer fringe of things in Vietnam because the usual military organization and efficiency that I had encountered so far was totally absent. There was no one to meet us nor were there any signs telling us what to do. We found a phone somewhere and made some calls and eventually a truck showed up to pick up those of us assigned to the 196th Light Infantry Brigade. It was here that I bid farewell to my Cajun helicopter pilot friend, who was assigned to the MEDEVAC unit.

"Well, take care of yourself and keep your head down."

"You do the same. If you ever get into any shit and need some help, I'll come for you. My call sign is 'Ragin' Cajun 3'"

The 196th Light Infantry Brigade, nicknamed "The Chargers" deployed to Vietnam in 1966 as a separate brigade and distinguished itself in fighting in three Corps zones (I, II & III). In 1968, it was incorporated into the newly formed Americal Division as one of its three infantry brigades. When the Americal Division was deactivated in 1971, the 196th again became a separate brigade made up of three infantry battalions (2/1, 3/21 & 1/46) an artillery battalion (1/82) and a cavalry squadron (1/1 Cav). I was beginning to hear that it was a pretty good outfit. If I couldn't be in my dream unit, being in a good unit was good enough.

Brigade Headquarters was located a few miles from the city of Da Nang at a place called Freedom Hill, nestled at the base of a hill mass between two villages and several rice paddies. Strategically, it was positioned to protect the air base and the city by sitting between them and the mountains to the west, from which any serious attack must come.

Finally, a deuce and a half truck arrived. I shouldered my kit bag once again and climbed in the back of the truck for the next leg of the journey. We were delivered to the personnel office at the brigade headquarters at which time the officers were separated from the enlisted men. I handed my orders and personnel file to the non-commissioned officer on duty. He said, "Would any of you gentlemen be interested in a job at Brigade Headquarters?"

I was suddenly gripped with fear at the prospect of being siphoned off to a rear echelon job where I would fight my war shuffling papers. But, we were being asked, not told.

"No way! I want to be assigned to an infantry battalion."

To my relief, my wish was granted. Normally, we would have been sent straight to the in-country familiarization school, but, since it was so close to Christmas, it was decided to send us to our units so that we would have a "home" for the holiday.

I was assigned to the 1st Battalion, 46th Infantry, nickname The Professionals, at Camp Crescenz. Each battalion had their own self-contained base camp with its own defensive perimeter. Camp Crescenz was separated from the Brigade Headquarters and the other battalions by a mile or so in the direction of the air base. Between us and the air base was a squalid village known locally as Dogpatch. The insignia of the 1/46 Inf. was a blue shield with a big torch on it. I thought it resembled a sign for a gas station. So the joke became that we worked for Professional Gas, a subsidiary of Charger Petroleum. We also jokingly called our brigade "The Burning Worm" a reference to the burning fuse on our shoulder patch. I was delivered to the 1/46 Inf. that night and spent my first night with my unit on Thursday, 23 December 1971.

4 EARLY DAYS

I had arrived at my battalion home but had no assignment down to company level as yet. I was housed in the lieutenant's hooch, which was basically a raised wooden hut with bunks down either side. It accommodated about sixteen. I discovered that I knew some of the other lieutenants in the battalion so that night I joined a couple of them at a floor show in the Non-commissioned officer's club, which was yet another wooden hut. The show featured a Korean group: young ladies decked out in the most garish Western style. Gigantic beehive hairdos, sequined mini-skirts, white go-go boots and crooning American favorites in pidgin English. "To dweam the impossible dweam."

We headed back to our hooch at about 2230 and, never having been much of a night owl, I went on to bed. At about 0200, I was abruptly awakened.

"Get your ass out of bed!! What the hell is going on here!?"

I arose to find the battalion Executive Officer, Major Lenhart, conducting a shakedown inspection of the hut. All the other lieutenants were standing loosely at attention in various states of undress. Maj. Lenhart was built like a football lineman, with forearms like hairy bowling pins, and he was bellowing mad about something.

"Goddamn it, I said get your ass out of bed!"

I looked in the direction that the Major was screaming and saw that there was somebody else in bed under the covers. Hesitantly, the covers came down to reveal a lieutenant, fully clothed and, lying next to him in the bed, an AK-47.

Now I was wondering what the hell was going on here.

"This is exactly why it's against regulations to keep weapons and ammo in the rear area. Playing with hand grenades. You stupid fuckers could have gotten somebody killed. Look at what would have happened if someone had been sleeping in this bunk. You're a sorry ass excuse for officers."

I looked at the bunk that he indicated and saw that there were holes where shrapnel had come up through the floor, through the mattress and ended up in the walls and ceiling.

It was beginning to dawn on me that someone had been playing with a hand grenade, accidently pulled the pin and then, in panic, thrown it under the hooch. It was also dawning on me that I didn't like Major Lenhart very much. He was treating all of us as if we had been active co-conspirators in what was admittedly a very stupid stunt.

I have no idea how I slept through the blast. The time was coming very soon when the slightest, imperceptible sound would awaken me.

It was a very memorable second night in my unit.

My Christmas in Vietnam was probably the worst that I ever spent. I had always envisioned Christmas in a war zone as being a sentimental experience, with its deep connection to family and religious tradition. Certainly the movies often portrayed it that way, with soldiers in foxholes or trenches singing 'Silent Night'. What made it horrible for me was that it was just another day, pretty devoid of sentimentality or even acknowledgement. I did attend a chapel service but that was so unmemorable that I have no recollection of it other than that I went.

I did have one present to open. Prior to leaving for Vietnam I met my cousin Fred Wellington for a beer at LaVals Gardens in Berkeley. He gave me a present and made me promise not to open it until

Christmas day. I kept my promise. The gift was a book of E.E. Cummings poetry, which I still have.

Fred had joined the paratroops in 1945 right at the end of WWII. He was barely of age and his eyesight was so bad that he would have been disqualified from enlistment. He memorized the eye chart and passed the test. He was too late for the fighting but served on occupation duty in Japan and later in the 82d Airborne Division. I, too, had been a paratrooper in the 82d so we had that in common and our visit at La Vals was very upbeat and full of optimism and encouragement. It was only many years later that Fred told me how opposed to the war he was and how angry he was that I was being sent to it.

I formally reported for duty within the next day or so. The Battalion Commander was Lieutenant Colonel Tate and he seemed to be pretty squared away. Part of the process of reporting in was to receive a briefing on the mission of the unit and who our opponents were. Our mission was to patrol what was known as the "Rocket Belt" in order to prevent the Da Nang Air Base from being subjected to a rocket attack. The rockets had a considerable range, so patrolling the "Rocket Belt" meant pushing well out into the Annamite mountains to the west of Da Nang.

The brigade AO (Area of Operations) was broken into two sub-areas: AO Maude and AO Linda with a firebase of the same name in each to provide fire support and command and control. A battalion was committed to each AO at all times. The third infantry battalion would man the ridgeline and perform other duties as required.

Charlie Ridge was located in AO Linda. This ridge was where most of the serious fights with the NVA had occurred and no one was anxious to go there. My platoon had been in a big fight there two months previously and the Platoon Sergeant had been killed. Our enemy consisted of the 575th NVA Rocket Battalion and the Q84th VC Company. Captain "Crunch" and I were taken up to the Ridgeline where we could look out over the areas described above and then we were given a close-up look via a VR (Visual Reconnaissance) in a LOH (Light Observation Helicopter). The LOH had no doors on it and the pilot provided thrills by abruptly turning the bird on its side so that I was looking straight down at the ground, held in by centrifugal force and the seat belt.

I was assigned to Company C, which was commanded by Captain William Thomas. On arrival there, I was further assigned to take over the 1st Platoon, which had previously been led by Lt. Jim Cole, who was being re-assigned as the Company Executive Officer. He gave me a description of the members of the platoon and I felt confident that I was getting a good bunch of soldiers. Now if only I could uphold my end of the bargain. Taking over a military unit is daunting at all times, but especially so for young officers who have yet to develop full confidence in their own abilities.

When the time came for me to formally take responsibility for the platoon in formation, I was posted at the rear of the platoon, waiting for my signal to come forward. Sergeant First Class Ferguson was in front, facing the formation and taking the report from the squad leaders. When that was complete he executed an About Face. This was my cue. I marched to the front of the platoon and came to a halt facing SFC Ferguson. He saluted me and said:

"Sir, 1st Platoon is present or accounted for."

I returned the salute and, in my best command voice, said "Post", at which the platoon sergeant marched to the rear of the formation. The platoon, and the lives of thirty men were now my responsibility.

"At Ease," I commanded.

"I'm Lieutenant Sherwood and I am your new platoon leader. I promise you my best effort and my support and I ask the same of you in return. If we work together we can get through this in the best way possible. I know that none of you want to be the last man killed in

Vietnam. Our best chance of staying alive is to stay alert, stay focused, and work together. PLATOON, A tench_HUT!…Platoon Sergeant!"

The Platoon Sergeant marched back to the front of the platoon and faced me.

"Take charge of the platoon."

We exchanged salutes and I executed an about face and marched away. Soldiers in formation are pretty inscrutable and it was impossible to read in their faces how I was received. At least I perceived no negative vibes.

FIRST PATROL

On the 30th of December, I was ordered to take out a squad-size patrol along a ridgeline to the north of our base camp. We would be taken by truck to the start point and I was given a map location and approximate time when we would be picked up. We were to look for any sign of enemy activity, check the papers of any civilians we encountered, and report accordingly. We were to engage by fire only if fired upon. At the drop-off point we were immediately faced with a steep climb to get up on the ridge.

I was no stranger to humping steep hills. A year previous, while in the Mountain Phase of Ranger School, I had the experience of humping up the mountains of northern Georgia with a heavily loaded rucksack so what we were about to do on this patrol was not a completely new sensation. However, in the intervening year I had smoked a lot of cigarettes and drank a lot of beer, especially on my 30-day leave before coming to Vietnam. The long and the short of it was that I was not in very good physical shape.

We began our ascent to the top of the ridge and it wasn't long before my heart was pounding and my lungs were heaving. I was acutely aware of my role as the leader and the requirement that I set the example in all things. My mind knew that but my body didn't and it was just by force of will that I kept going. Every time we reached the "top" where I hoped the going would be easier, we were greeted with yet another increase in elevation. At the top of one hill, we encountered some older Vietnamese women and men cutting wood at the bottom of the other side of the hill. The Kit Carson Scout (KCS) who was with us called out to them, "Lai day, Lai day."

They effortlessly scampered up the hill in their bare feet and submitted to questioning by the KCS. "Have you seen any VC?" "No BC, no BC." Completely un-winded, one ancient mama-san immediately squatted down and lit up a black, gnarled stogie. Meanwhile I was still gasping for breath and tasting blood in the back of my throat. It was quite a lesson in humility for this rough-tough American fighting man. After checking papers and hearing the "No VC, no VC" mantra, we proceeded on our way across the rugged hill mass.

"LT! There's two dinks on that hill over there."

About 150 yards away, there were two Vietnamese men who appeared to be digging. When they saw us they took off, carrying what looked like artillery or mortar rounds. Unexploded artillery or mortar rounds in the hands of the VC were the makings of a very deadly booby trap and their activity looked anything but benign. They were well within rifle range but they hadn't fired at us and we couldn't be positive at this distance that they were up to no good. Grabbing the hand set for the radio, I called our higher headquarters.

"Lima zero three, this is Lima one two over."

"Roger one-two, this is zero-three, over."

"Ahh, we have two Vietnamese males carrying what looks like artillery or mortar rounds. They are approximately 150 meters away to our November Echo on another terrain feature. Request a bird to this location for a closer look and intercept, over."

"Roger. What is your location?"

Quickly I determined the grid coordinates of our location and then converted it to code, using the encode/decode book that all leaders and radio operators carried.

"I'm at, I SET, Alpha Papa — Foxtrot-Romeo-Delta-Yankee-Charlie-Whiskey, over."

"Roger, I copy Alpha Papa — Foxtrot-Romeo-Delta-Yankee-Charlie-Whiskey , over."

"Good copy, over."

"Stand by, over."

"Standing by, out."

"Hey, they're hatt'n up."

"LT, the dinks are di di-ing."

"Lima zero-three, this is Lima one-two, over."

"This is zero-three, go."

"Roger. We need that bird ASAP. Suspects are moving off to the Sierra Echo, over."

"Ahhh, this is zero-three, roger, stand by."

I stood by — and stood by. Instead of a prompt response, I got the run around — not unlike telephoning the police for assistance and being put on hold or being told by a computerized voice to negotiate a menu of services. Meanwhile, our two suspicious characters weren't waiting around for us. They disappeared over the horizon. "Shit!" I was pissed. My anger was only in part due to the lax response to what was clearly a suspicious activity. The other part was that I had been denied the opportunity to score a coup on my first mission — and what a coup that would have been.

The remainder of the patrol passed uneventfully except for finding a cave that had not been used recently. In the afternoon, we arrived at the pre-arranged pick-up point and I radioed for the truck to come and get us. When the deuce and a half arrived, we happily flopped in the back for the ride back to our base camp. Back at camp the squad and I popped a beer and discussed the day.

In tallying up the day, it wasn't all bad. On the negative side, we had failed to get the two Vietnamese men, who had been just outside our reach. Also, I was in really bad shape, physically. On the plus side, the squad members were competent and had responded well to me. For my own part, I had navigated successfully and not gotten us lost. And, in spite of my being out of shape, I gutted through it. All in all, not a bad day.

5. PREPARATION

The battalion was scheduled to begin a forty-four day mission in AO Maude on the 8th of January, so the first week of January was given over to preparation for this extended period of combat operations in the field. It was also a week for me to learn the lay of the land, sometimes in unpleasant ways.

Equipment had to be checked, weapons zeroed, orders for the mission received and transmitted down to the squad and soldier level so that each man knew what was going on and what was expected of him. I was a very lucky lieutenant to have this time to prepare and, more importantly, to get to know my men and allow them to get to know me. I knew guys who had been flown out in the middle of a battle to replace a platoon leader who had just been killed.

One day we were taken to a rifle range to zero our weapons, which is a process of adjusting the sights of the rifle to match the eye of the shooter. The company was piled in three deuce-and-a-half trucks, with my truck at the head of the small convoy. As the senior occupant of the truck, I was riding in the cab. In order to get to the rifle range, we had to drive through Da Nang. This was the first time I had been in the city. All of a sudden, we came upon a protest demonstration and the road was blocked off with a barricade of concertina barbed wire. I got out of the truck and approached an older gentleman who seemed to be the leader.

"What's the problem here?"

Papa-san was very agitated and responded in a stream of Vietnamese. It was immediately clear to me that I probably spoke more Vietnamese than he did English and I only knew a few words and phrases, none of which were going to be very helpful in this situation.

"We need to get through."

Another tirade in Vietnamese.

"What are you going to do now, Lieutenant?" I thought.

As I was pondering my next move, the platoon sergeant from one of the other platoons appeared.

"Goddamnit, get this fuckin' shit out of here." he said, as he grabbed the concertina barricade and dragged it aside so that we could proceed.

It was obviously the correct move and I felt a little stupid and ineffectual that I had not been as immediately decisive as the sergeant. I also had another feeling and that was embarrassment that we had been so disrespectful of the Vietnamese. Whatever they were protesting, it was clearly very important to them and we had brushed them aside as if they were misbehaving children. You do what you have to do, but sometimes it doesn't feel good.

We continued on our way. The zeroing of the weapons and the return trip were uneventful.

On another day, the company was ordered to conduct a CA (combat assault by helicopter) onto Da Nang Air Base. This was a practice of a contingency mission that the brigade had to reinforce the air base in the event it came under direct attack, something that had happened before.

The immediate task on a CA is to form a hasty defensive perimeter until the whole unit is on the ground. The battalion helipad could accommodate five Hueys at a time, enough to lift one platoon. Each of the three platoons would be assigned a sector of the perimeter using the clock system, the direction of flight marking 12 o'clock. One platoon would occupy from 12:00 to 4:00, the next from 4:00 to 8:00 and the last from 8:00 to 12:00. All of us, myself included, were well trained and practiced in airmobile movement and the practice mission went smoothly. The battalion commander, Lt Col Tate, was on the ground to observe and he seemed pleased. At the conclusion of the exercise, he gathered the company around him to give us a pep talk.

"Well done, Chargin' Charlie! Charlie Company is my strongest company. I know that if the shit hits the fan, I will be able to count on you guys to get the job done."

To be so congratulated by no less a demigod than the battalion commander pleased us no end.

Around the base camp, I was beginning to note evidence of the racial tension that had been so publicized by the stateside media. Virtually all the Black soldiers performed a ritual handshake known as the 'dap'. There was the basic dap, which I had been taught by some of my soldiers, but there was also several elaborate variations, each with a name such as 'Power to the people'. I had no problem with this symbol of racial solidarity within the parameters of good order and discipline.

There were, however, some angry, militant Black soldiers who would use the dap as a way of showing their disdain for the Army, and officers in particular. One soldier in particular stands out in my memory because of his Louisiana name and his hostile attitude. He seemed to be the center of gravity for Black militancy in Charlie Company. His name was Schexnayder but he went by 'Bro Schex'. If an officer was approaching, Schexnayder would make it a point to get involved in the dap ritual and ignore the officer by not rendering the customary salute that is passed between officers and enlisted men.

One day Captain Thomas, who was my company commander and was black, walked past Bro Schex and didn't receive a salute. Captain Thomas lit him up like a Christmas tree. It wasn't so easy for us young, white lieutenants so we tended to avoid getting stuck in that predicament. I was glad that Schexnayder was not in my platoon. The Black soldiers in my platoon were from small Southern towns like Vicksburg, Mississippi, Bastrop, Louisiana, Swainsboro, Georgia, and Spartanburg, South Carolina. They were good guys who were liked and respected by everyone. People tended to self-segregate by race in social situations, which to me was fairly natural, but in the field, everyone worked together.

One morning, Captain Thomas called the three platoon leaders together in his office and spread a map out on the table.

"We're going in here," he said as he placed a sausage-size finger on the map, covering a one square kilometer grid square.

"We'll move up into the high ground here and patrol up along this ridge."

The sausage moved along great swaths of terrain.

All my training and experience up until now had emphasized the need for precision when pointing out features and routes of movement on a map. In Ranger School, we were taught to use a pine needle or something similar when briefing anyone on a map. I assumed that this was a Warning Order, a way of alerting us of the upcoming mission prior to the issuance of the much more detailed Operations Order. Following the finger waving exercise, Captain Thomas said, "Are there any questions?"

"When will you be issuing the Operations Order?" I asked.

"You just got it."

This was, I suppose, the beginning of a less than harmonious relationship between Captain Thomas and myself.

I cobbled together as much information as I could and prepared my own platoon Operations Order to issue to my squad leaders. But where to issue it? I hadn't been in the battalion long enough to know all the nooks and crannies where one might gather for a meeting and a little privacy. The one place that offered what I was looking for was the

Officer's Club, as the wooden hut with screen siding was euphemistically called. There was the main room with a bar but in the back was a smaller room where movies were sometimes shown. No one was ever in the 'O' Club in the morning so the back room seemed like the perfect venue for briefing my squad leaders.

I met them at the company headquarters and we walked down to the 'O' Club together and I issued them my order and answered their questions as best I could. When there were no more questions I dismissed them to go and prepare their own orders to give to their soldiers.

Later that day I was approached by a captain from the battalion staff.

"Lieutenant!"

"Yes, sir?"

"It's been reported to me that you had enlisted men in the Officer's Club. Is that true?"

"Yes, sir, I had my squad leaders in there to issue them an Operations Order. There was nobody in there."

"Lieutenant, the Officer's Club is off-limits to enlisted personnel and you shouldn't be fraternizing with enlisted men."

"Sir, I wasn't fraternizing, I was issuing an order."

"Lieutenant, do I make myself clear?"

"Yes, sir."

I was too immature to not let this anger me but there was nothing more to do about it so I focused on continuing to get myself and my platoon ready for the upcoming operation in AO Maude.

6. 8 January 1972

8 January 1972. This was the big day. We were going to the field for a forty-four-day mission in the Maude AO. We were up well before dawn to draw weapons, ammunition, and rations from the supply room and get a hot breakfast. It was raining and, as the day dawned, we were enshrouded in dense fog. After breakfast, we made our way to the chopper pad to await the helicopters that would take us to the field where we would relieve the 2d Battalion, 1st Infantry, who would be returning on the same birds.

I broke the platoon into five chopper loads and positioned them at the five points on the pad where the birds would land. Then word came down that the birds were socked in by the fog on their base at Marble Mountain and that they would pick us up as soon as the weather cleared enough for them to take off. This could happen at any time so we sat in the rain and waited. At about 1100, I started a letter to my parents but was only a few lines into it when we were told to break for lunch. No sooner had we gotten our meals than we were told that the weather had cleared and that the birds were on the way. In a rush, we made our way back down to the pad and resumed our pickup positions.

Before we could see them, we heard the characteristic, never to be forgotten 'Wup, wup, wup' of the main rotor as they approached. At each pickup point, a soldier raised his hands as in signaling a touchdown to mark the spot where the chopper was to land. In came the birds, nose up for landing and lashing us with rotor wash and rain. As soon as the skids hit the steel planking of the pad, the six soldiers at each pickup point boarded, three to each side, sitting on the floor, legs hanging out of the door-less aircraft.

Once everyone was aboard, the pilots revved the engines for take-off and we were on our way. The flight pattern took us south in the direction of Da Nang and we could see the huge white statue of the Buddha in the distance. Then we swung east in the direction of the South China Sea, only to continue the swing 180 degrees until we were heading west toward the mountains. It was a little unsettling sitting there with no safety restraint and only a few short inches from the edge—especially when the helicopter would bank and we would be looking straight down into oblivion.

The terrain we were passing over between Da Nang and the mountains was fertile farmland—a checkerboard of rice paddies with villages interspersed here and there. Much of it was now abandoned, the land a green wasteland pockmarked with artillery and bomb craters and destroyed villages. For the United States, this war was now old and tired and tottering toward its conclusion. How many hundreds of times before me had soldiers flown this same route and hunted an elusive enemy in these same glowering grey-green mountains? But for me, it was new and fresh and I was full of excitement as I hurtled into battle in my airborne chariot.

After flying for several minutes, I spotted violet smoke billowing in a field of bright green near the base of the dark, mist shrouded mountains. This was our landing zone. As we got closer, I could see a soldier standing behind the smoke with his rifle raised over his head, indicating where the lead aircraft should land.

As we got off the birds and formed our hasty perimeter, an equal number of soldiers from the unit we were replacing got on for the return trip to their base camp.

It was raining and cold as we lay soaking wet and waiting until the whole company was shuttled to the field and assembled. Once that was complete, Captain Thomas gave each platoon a direction of march, with general instructions to look for signs of enemy activity.

I started moving off the LZ to the north when all of a sudden I had a stabbing pain in my left ankle. Shock, disbelief and blinding pain. I looked down to find the source and discovered that I had stepped into a hole just big enough for my foot to fit into. All 170 pounds of me had come down on my ankle. I quickly pulled my foot out of the hole as if by doing so I could reverse the situation and undo what had happened. We were on the move and I couldn't stop nor did I feel it appropriate to even show that I was in pain, so on I humped.

We were in the low foothills on the western edge of the mountains and the terrain was fairly open — lightly wooded and high grass. In the late afternoon, we came to a small hill which would just barely accommodate the platoon for our Night Defensive Position (NDP). It wasn't ideal, but it was the only defensible terrain around. I assigned each squad their sector of the perimeter and then the platoon sergeant, Sergeant First Class Ray Ferguson, and I went around checking each position, placing the machine guns and making sure that dead-space was covered by the grenadiers. By the time I had gotten the platoon situated, I started looking for a position for myself. All that was left on this little hillock was two saplings far enough apart for my hammock. I strung my hammock and then strung my poncho over that, creating a shelter from the rain.

Next I dispatched a squad to patrol the area around us. After they had been gone for a while a call came on the radio. In a hoarse whisper, I heard:

"One-six this is zero-three, over."

"This is one-six, go," I responded.

"Roger, we are hearing movement and chopping noises, over."

"This is one-six, Roger. Can you get a direction and distance, over?"

"Ahh, this is zero-three, direction from my location is about 350 degrees. Distance is hard to estimate. Maybe about thirty to forty meters, over."

"Zero-three, this is one-six. Send me your location and hold in place. I'm going to bring another squad and link up with you, over."

"This is zero-three, WILCO, over."

"This is one-six, out."

"Sergeant Ferguson," I called.

"Yes, sir."

"Tell Sergeant Carlson to saddle his squad up. As soon as I get a location on Third squad, we're going to link up with them and check out the noise they're hearing. You stay here with Second squad and secure the perimeter."

"Right, sir."

Once Third squad had sent in their location and Sergeant Carlson's squad was ready, we moved out.

Third was no more than a kilometer away and the link up was smooth. Once together we started patrolling in the direction of the noise they had heard. All we found were several large grass piles, hollowed out in the middle in such a way that it could have accommodated a human. Other than that, nothing. If anyone had been there, they probably di di 'ed* (di di mau—Vietnamese for 'leave') the area when they heard the squad leader on the radio. Thus began a seemingly endless game of cat and mouse, looking for the elusive Charlie, who always seemed to have disappeared just around the corner.

Back at the NDP, more checking of positions and making sure things were set for the night.

Finally, it was time to get a little shut-eye. Climbing into my hammock, both saplings bent in to my weight, so that I was reclining in a 'V' position. I had yet to learn the little tricks of setting up a hooch, such as tying a tuft of grass on the hammock strings to wick away the rainwater, which would naturally flow down them. Consequently, I spent the night in a 'V' position, with my ankle throbbing and cold rainwater pooling on my ass. It was a most uncomfortable night, to put it mildly.

I had hoped that a night of rest would be all my ankle needed, but when I tried to stand up in the morning, I fell down. I couldn't stand on it, let alone walk on it. I was face to face with the fact that I was going to have to be evacuated on my second day in the field. Furthermore, I was now afraid that my ankle was broken, having never before experienced a sprain that I couldn't deal with.

After consulting with Captain Thomas over the radio, a dust-off bird was called to evacuate me. The bird arrived shortly after it was called for, landing on a flat area at the bottom of our little hill. A couple of my soldiers helped me down the hill and got me situated on the aircraft. As the bird lifted off and I waved to the platoon, I wondered if I would ever re-join them and, if I did, would they accept me after this ignominious beginning.

7. BOOBY TRAP

The bird delivered me to the battalion aid station, which was just off the edge of the chopper pad. I hobbled in, using my rifle for support, and explained to the medic what had happened. After doing an examination of my ankle, he got it x-rayed. I sat there in dread waiting for the results of the x-ray because I was sure that my ankle was broken and that my days as a field soldier would be over. After a while, the medic came back and told me that it wasn't broken, that I had a bad sprain and that I would have three days light duty in the base camp and that I could re-join my platoon after that. Immensely relieved, I made my painful way to the lieutenant's hooch.

After three uneventful days kicking around the base camp, I caught a bird back to the field to re-join my platoon. I was still in pain and walked with a limp but I was functional. I knew that I needed to put on a brave, confident face, but inwardly I was worried that I had lost face and that the platoon would not accept me as their leader. As I entered the platoon perimeter I was greeted with a nonchalant, "Hey, LT" and I realized that my big deal was not their big deal. The general response to my return was as though I had never been gone. Sergeant Ferguson brought me up to speed on the previous three day's activities and I seamlessly resumed my duties.

Not long after re-joining the platoon, we were moving West through the foothills on our way up into the mountains. I usually walked third in line behind the point man and the slack man. The point man was the eyes and ears of the platoon and it was important to be positioned where I could see him and be alert and responsive to him. Each of us also had to be alert to our own surroundings, continuously looking to the front and flanks, up and down.

Suddenly the point man raised his fist and crouched down. I immediately followed suit, crouching down and giving the same fist sign, which would be repeated down the line until everyone had halted and faced out, prepared for danger. After carefully looking around his immediate area, the point man came back to me.

"Sir, there's a booby trap."

He pointed to the left front to a spot parallel to his furthest advance. Sure enough, I could see a 105mm round tied to a tree at about shoulder height. The silence was absolute.

I got on the radio and called the platoon sergeant, who was positioned to the rear of the platoon.

"One-five, this is One-zero, over."

"This is One-five, over."

"This is One-six, we've got a booby trap here. Move everyone back and take cover. Send the demo man up, first of last 'Mike'."

"This is One-five, roger that."

Now there was some noise as the platoon began to move back in the direction from which we had come. Momentarily, burley, barrel-shaped McCartney appeared. He loved explosives and was the self-appointed demolitions man for the platoon.

Pointing, I said, "McCartney, we've got a 105-round tied to a tree over there. Do you see it?"

"Yes, sir."

"Good. OK, what I need you to do is make sure that the triggering device is deactivated. Once that's done, get it off the tree and into a hole so we can blow it in place. Got it?"

"No sweat, LT," McCartney said as he scuttled off in the direction of the booby trap.

While McCartney went about his business, I encoded our location and called in a situation report to Captain Thomas.

"Charlie-six, this is one-zero, over."

"This is Charlie-six Alpha, over." (Charlie-six Alpha was the CO's radio operator (RTO)).

"Roger, Charlie-six Alpha. We've encountered a booby trap at, I set: Whiskey Sierra, Quebec, Tango, Foxtrot, Yankee, Bravo, Mike. We are in the process of deactivating it and will blow it in place, over."

"Good copy. Give a heads up when you get ready to blow it, over."

"WILCO, out."

A few minutes later, McCartney returned.

"OK, LT, I've got it in a shell hole wrapped in Det-cord* (*Detonation cord. Explosive cord resembling plastic-coated clothes line*) and ready to blow it, when you are."

"OK, good. Let me call Sergeant Ferguson to get the platoon ready and let the old man know. I'll give you a thumbs-up when we're ready. Make sure and give the warning before you blow it."

I called Sergeant Ferguson and told him to get the platoon down and then called the CO to alert him that we were getting ready to blow. I gave McCartney a thumbs-up and then took cover in a bomb crater with Olsen, Bishop, and Granado. There was a pause while McCartney lit the fuse leading to the blasting cap at the end of the det-cord, then dashed for cover, bellowing the standard warning:

"Fire in the hole, fire in the hole, fire in the hole!"

KABOOM

Rocks and debris flew through the air and a big cloud of gray smoke rose from the hole. Then, silence, as if nothing had happened.

I sat with Olsen, Bishop, and Granado for a few more minutes in the bomb crater and had a smoke and then gave the signal to saddle up and resume our march into the mountains. No drama, not even any suspense music in the background. It was as if nothing out of the ordinary had happened.

SP4 Stephen Olson

PFC Cecil Dodd

Grenadiers-- SP4 Jesse "Blooper" Brown and SP4 Ruben Bugge

8. MOUNTAINS

The semi-open foothills gave way to thick undergrowth as we began moving up into the mountains. In spite of the double canopy forest, enough sunlight penetrated to produce thick vegetation at ground level, which made uphill travel a real grind. Our aim was to get up to the top where movement along the ridge would be relatively easier, but in order to get there, we had to hack our way straight up the side. At the front of our file, the point man, with 90-plus pounds on his back, rifle in one hand and machete in the other, hacked a path through the vines and brush while the rest of us staggered along behind. Cutting point was exhausting work and point men had to be switched out at frequent intervals. We were a platoon, but, at the moment, we were thirty individuals slogging uphill through the red clay mud, alone with our thoughts.

God damn, this is a bitch.

This mud is unbelievable. My feet feel like they weigh ten pounds and I can't get a foothold to save my life.

I reach out and grab a sapling to my left front and haul myself up a step. My ankle is throbbing.

I take another step and my foot slips back and I go down on my knees. I dig my toes in and use the butt of my rifle to steady myself as I haul myself back up. I lean forward and shrug the weight of the rucksack off my shoulders for just a delicious moment. Then I take another step, grabbing at a bush for support, while the full weight of the ruck digs into my shoulders once again.

I take another step and am unable to move. Something is holding me back. A jungle vine has caught on my rucksack. My anger flares and my first instinct is to try and bull my way forward. But then I remember the words of one of our instructors in Ranger School, "Don't try to fight the jungle, you can't win. Work with it." I turn to the right, easing the tension so that I can slip out of the bind and move forward again.

This pattern repeats itself for about 45 minutes. My chest is heaving, my lungs are about to burst, and my legs feel like they are made of rubber. I don't think I can take another step.

I whistle softly to the man in front to get his attention. I raise my fist to signal 'Halt' and then spread my fingers to indicate a five-minute break. I turn around and give the same signal to those behind. Once satisfied that everyone has stopped, I turn around, facing downhill, and flop down, leaning against my ruck on the uphill slope. I brace my feet against a small tree to prevent myself from sliding back down. I'm drenched with rain and sweat. Reaching for the towel which is always draped around my neck, I wipe my face. It stinks of ammonia and other bodily excretions. I fumble in my jacket pocket and find the plastic box in which I keep my smokes, extract one and light up. I'm still breathing heavily but not so much so that I can't take a deep drag on my cigarette. I look back down the hill at the backs of the heads of my comrades, spaced evenly five meters apart. Wisps of smoke from other cigarettes float lazily upward in the rain soaked air.

'God damn, this is a bitch. I wonder how much further to the top? Is this killing them like it is me?'

It takes about five minutes to smoke a cigarette and I know it is time to move again. I don't really want to, but I know we have to. Again, I whistle softly to those ahead and twirl my index finger in the air, indicating that it is time to move. I turn downhill and do the same. There is a rustling as everyone struggles to their feet. We resume the march. We are grunts.

Finally, in the afternoon we hit the ridge. Not surprisingly, there is a trail on it and not much choice but to use it. It is too late in the day to do much patrolling so I select a spot on the trail to set up for the night. The slopes off the ridge are too steep for encampment so we set up a long, narrow, "Cigar shaped" perimeter astride the trail. I send patrols up and down the trail to make sure there is no unwelcome company and then set up mechanical ambushes (MA) on the main trail in both directions, plus one on the trail we made coming up. It feels wonderful to be out from under the weight of the ruck.

I encode our position and have Cameron, my RTO, call it in to the company CP (Command Post). While he's doing that, I check the perimeter to make sure that everyone has cover and concealment and that weapons are properly positioned. Foliage needs to be cleared away so that everyone has a field of fire to their front.

"Sergeant Rodriguez, shift Hanisco's sixty a little to the left so that he can sweep the whole trail with it. Where he's got it now, he can only cover the trail right in front of him."

Hasty fighting positions also need to be dug. Satisfied that work is progressing and that Platoon Sergeant Ferguson is supervising, I return to my position, pull out my map, and begin plotting Delta Tangos (Defensive Targets) around our perimeter as well as targets along tomorrow's route of march. Once these are called in and registered with the FO (artillery Forward Observer), they can be called for very quickly in the event of an emergency. Our FO is nicknamed 'Tee Tee LT', owing to his diminutive size. He is a cool guy and everyone likes and trusts him.

Again I inspect the perimeter, checking on the progress of digging foxholes.

"Cooke, what do you call that?"

"It's my fighting position, sir."

"You couldn't even fit your ass in that. I'm only requiring prone fighting positions for tonight, so get with it!"

"Yes, sir."

The rain has let up as darkness begins to descend. The air turns to a heavy mist and we remain soaked through and through. We prepare our supper before it gets totally dark. Everyone has their own favorite meal. Mine is C-ration Beef with Spiced Sauce over the Korean Top-ramen noodles with which my hooch maid supplies me. Tastes vary but there is one C-ration meal that is universally hated and that is Ham and Lima Beans or, as we call them, Ham and Mother Fuckers. The acrid smell of trioxin heat tablets wafts around the perimeter as the guys heat their meals over their 'Ranger stoves'. (A small C-ration can with air holes cut in the sides and in which is burned a heat tab). You learn very quickly never to try cooking with heat tabs inside a hooch. The fumes are worse than tear gas.

It is completely dark now and Platoon Sergeant Ferguson has set up the radio watch for the night. There has been no sign of enemy activity so we are on thirty percent alert.

Sergeant Ferguson and I talk quietly in the dark.

"Company wants us to submit the name of someone to be supply clerk. You have any recommendations?"

"I'd say Baker, sir. He's been in-country about ten months and he's a good man. He's earned a break."

"Baker is a good soldier and I don't want to lose him out here. But you are right, he deserves a break if we can give him one. I'll call his name in. Besides, he might be able to cut us a break every now and then as supply clerk."

After chatting for a few more minutes, I decide to turn in and get some sleep. From the waterproof bag in my ruck I pull out my nylon blanket. I loosen the laces of my boots but keep them on, remembering the admonition of the Ranger instructor, "Rangers, never take your boots off at night. Your shit will be weak if Charlie hits you at night and you're fucking around trying to put your boots on."

I cradle my rifle in my arms and wrap up in the blanket. I am still wet but the blanket helps to hold in my body heat. I sleep lightly, part of my brain hearing everything and listening for any sign of trouble. Next to me, the radio watch keeps his silent vigil. Periodically I hear the slight rustle as the watch is changed, with the new one arriving to take his watch and the other going off to get some sleep. Every hour I hear the radio static and the hourly situation report.

Kshhhhhht "Charlie six, this is Charlie one zero, over." Kshhhhhhht.

Kshhhhht "Charlie one zero, this is Charlie six, over." Kshhhht

Kshhhht "This is Charlie one zero, negative SITREP, over." Kshhhhht

Kshhhhht "Charlie six here, Roger, out." Kshhhhhht

At midnight, I make a mental note that the radio frequency is being changed. For security purposes, radio frequencies are changed every twenty-four hours, at midnight. Failure to make the switch would leave us out of touch with everyone.

This is what passes for sleep in the bush.

At 0500, the radio watch whispers to me that it is time for stand-to. I immediately sit up, tighten the laces to my boots, put on my web gear and helmet and take my position next to the radio. Cameron, my RTO, is also up and he takes over the radio from the night watch. All around the perimeter we are at one hundred per cent alert, waiting for the dawn. At dawn, the MA's (mechanical ambushes) will be brought in.

I have diarrhea much of the time and dawn finds me crouched next to the trail, entrenching tool and packet of bum wad in one hand, rifle in the other, waiting urgently for the MA to be brought in so that I can run down the trail and relieve myself. This frequently observed scene amuses Sergeant Rodriguez, who is nearby with Granado, saying with a chuckle, "El teniente tiene que cagar."

After my ablutions, I return at a more relaxed pace. Once back in the perimeter, I light a heat tab and start heating a canteen cup of water for coffee. Digging in my ruck, I extract a can of peaches which, along with a cup of coffee, will constitute my breakfast. At home, eating canned fruit would be a distant second to fresh fruit, but here it is a delicacy. I savor every bite and carefully spill the last drops of syrup into my mouth. I lick my plastic C-ration spoon "clean" and stick it in my left breast pocket. The coffee is instant, something else I would turn my nose up at under different circumstances, but here it's all we have. The bitter taste clears the palate and the caffeine gives me a jump-start for the day. Sergeants Ferguson, Carlson, Christiansen, and Rodriguez join me for coffee as we discuss the day's operation. It will have its own particularities, but for all intents and purposes, it will be a lot like yesterday, and tomorrow will be a lot like today.

9. COMINGS AND GOINGS

This entity of which I was in charge, with the outwardly unchanging title of First Platoon, Company C, was inwardly a living organism in a constant state of transition. This should not have been a surprise to me since I myself had arrived to replace someone, but the rapidity and constancy of the change came as something of a shock to me.

Very early in the Maude mission, I received a radio message requesting that I provide a soldier to become the supply room clerk. After consulting with Sergeant First Class Ferguson, I picked Specialist Baker. Baker was a solid soldier who had been in the field for about ten months and was nearing the end of his twelve-month tour in Vietnam. This was a way of rewarding him but it was also a loss of a reliable soldier from the platoon. Baker's appreciation for this gesture would come a couple of weeks later.

Then Melvin Smoots came to the end of his twelve-month tour and rotated back to the States. Sergeant Peasley was transferred to the mortar platoon. Our medic, "Doc" Cousino, was replaced by "Doc" Kroze. But the biggest blow of all came barely three weeks into the mission when Platoon Sergeant Ferguson and Staff Sergeant Carlson both came to the end of their twelve-month tour and rotated home. They were the only senior, experienced NCO's in the platoon, each having served one or two previous tours in Vietnam. The night before they left, I solicited parting words of wisdom from Sergeant Ferguson.

"So what advice would you give me before you go?"

"You're doing fine, LT. You're getting the hang of things and you'll be just fine. Here, I want you to have this. I won't be needing it any more."

He handed me his sheath knife, which looked like it had seen long service. I deeply appreciated the gesture and the vote of confidence.

"Who would you suggest to replace you?"

"Well, I'd recommend either Sergeant Christiansen or Sergeant Rodriguez. They both have their shit together."

Sergeant Christiansen had slightly more combat experience than Rodriguez and was a "hard striper", meaning he had come up through the ranks, unlike Sergeant Rodriguez, who was a "shake and bake", a product of the Non Commissioned Officer's Academy. On the other hand, Sergeant Christiansen had a volatile temper and could throw a tantrum when he felt slighted or didn't get his way. Sergeant Rodriguez was highly competent and much more even-tempered—an asset for someone who had to deal with the whole platoon. In the end, I chose Sergeant Rodriguez, moving Sergeant Stringfellow up to take over Rodriguez' squad.

New people were coming in on the same birds that took people out. One such was Jimmy O'Brien, soon to be simply known as "OB". He was the quintessential Irish kid that appears in every war movie. He had sandy red hair and a broad Irish face with fair, freckled skin and a ready smile. He was smart and eager to please, fitting so seamlessly into the platoon that it seemed as if he had always been there. Other replacements didn't fit in quite so smoothly, needing some breaking in before they figured things out.

One night, sitting in the middle of the perimeter with Sergeants Christiansen and Rodriguez, Sergeant Christiansen said:

"You know that new guy, De Mott? You wouldn't believe what he said to me this evening."

"What? What did he say?"

"He says to me, he says, 'Hey Sarge, what time are we going to get hit?'"

"You gotta be shittin' me. What did you say?"

"I said, 'I dunno, the dinks haven't told us yet. When they do, you'll be the first to know.'"

We laughed quietly at the strange ways of FNG's (Fuckin' New Guy), me conveniently forgetting that I was hardly an old salt, having been in the field for about three weeks. Having said that, the learning curve was steep and essential lessons had to be learned quickly.

On another night, there was a likely looking trail near our night position on which I decided to put an ambush. I left the platoon in the charge of Sergeant Rodriguez and took a squad out on an ambush patrol. We left the perimeter after dark and moved quietly to the ambush site adjacent to the trail. I placed each person in position, moving from right to left. After getting the left flank person in position, I moved to the center and took my place in the line. It was pitch black as we lay in wait for an enemy patrol to come down the trail and into our kill zone. All of a sudden, I heard a sound:

Whshhhhhhhh. Whshhhhhhhh

I thought, "What the fuck is that?"

I began low crawling as fast as I could down the back of the ambush line in the direction of the noise, which continued unabated.

Whshhhhhhhh. Whshhhhhhhh

As I closed in on the source of the noise, I discovered it was coming from one of the new replacements.

"What the fuck are you doing?" I hissed.

"I'm blowing up my air mattress, sir," he whispered.

"Knock it off. You don't need that and you're making a huge racket."[1]

"Yes, sir."

Although I couldn't see in the dark, I could almost detect a look of sulky disappointment, the look of a child who has failed to please a parent.

In the event, the ambush was a dry hole and perhaps that was a blessing, given that you never quite knew what an FNG would do.

And so it went, people leaving and new people coming. In spite of the constant changes, there remained a solid core of NCO's and men which maintained the integrity of the platoon in the midst of this seeming chaos. Christiansen, Rodriguez, Bishop, Elzey, Dodd, Granado, Salazar, Brown, Hanesco, Tetrault, Garcia, Bugge, Olsen, McCartney and Warner. These were the hard core which gave the platoon its personality and strength, and to which all new comers must adhere or be sloughed off.

SP4 Danny Bishop, author, PFC Ray Hanisco, SGT Christiansen, SGT Rodriguez, "Doc" Kroze, PFC Richard Garcia

---

1. The air mattress was the first item of equipment to be left behind in the base-camp. It was bulky, weighed a ton, and was of minimal use in the field.

10. HILLS, HILLS AND MORE HILLS

Hills, hills and more hills. Every time we got to what looked like the top of a hill, there would be yet more to climb. And if we had to go downhill for some reason, it only meant that we would have to go back up. Every day, the weather seemed to be getting steadily hotter, sapping our energy and making the climbing that much more difficult. My ankle was still quite sore and I walked with a limp.

One hot, steaming day, we were grinding up yet another hill, hacking our way up the nose of a finger leading up to a long ridge which looked down on the Song Lo Dong river to our north. We had filled our canteens in a stream at the bottom of the hill but our water was being rapidly depleted as we replaced the water that was being sweated out by the bucket full. Our fatigues were soaked through as if we had been standing in the rain. It was mid-day and we were about half way up when I spied to our right a level shelf of land sticking out from the side of the hill. It was about fifty yards by twenty-five yards and fairly open—a perfect size to accommodate the platoon for a lunch break. I raised my fist to signal a halt then got on the radio.

"One-one, One-two, One-three, this is One-zero, over."

Each squad answered in turn.

"Roger, we're going to halt in an open area to our right. The east end is twelve o'clock. One-one, take twelve to four, One-two, from four to eight, One-three, from eight to twelve. How copy, over?"

"Good copy, over."

"Roger, let's move. Out."

I accompanied the lead squad as they moved in and occupied their sector of the perimeter, using the clock method. Sergeant Rodriguez came up with the trail squad and together we supervised the positioning. Calling the squad leaders over, I informed them that we would move out in an hour and to stay alert. Rodriguez and I sat down in a central location.

"What's for lunch, LT?"

"Well, let's see what we've got in here."

Rooting into the pouch where I kept my rations, I came up with a can of turkey loaf.

"How's about some turkey loaf on crackers. Or, if you want it to be chicken, we can call it that. You got any crackers?"

"Yeah, here."

Each of us had a couple of crackers with a hunk of turkey loaf on it, chased by plenty of water.

"Damn", I said, "I wish we could catch these fuckers half-stepping. This is getting old just humping our asses off every day and never making contact. All that's happened so far is a couple of booby traps."

"The dinks are all on Charlie Ridge. I don't even want to go back there. We were up there back in October and had a hell of a firefight with the NVA. We were moving along a ridgeline and the Platoon Sergeant was walking point with Granado, when all of a sudden he yells 'Contact' and then dropped. They nailed him right between the eyes. Then all hell broke loose and Granado was stuck up there with Sergeant Thompson. The LT got a Bronze Star for it but it was Granado shoulda gotten it."

"Man, I definitely don't want anyone to get killed, but I sure would love to catch them red handed and light them up."

"There it is."

"I gotta take a leak."

I walked over to the edge of the clearing to relieve myself.

Damn my piss is yellow. I need to drink more water. ...What the hell is that!?... Fuck! Those are ticks—on my dick!!

"I got fucking ticks on my dick!!"

Word spread quickly and everyone started checking themselves out. Virtually everyone was covered with ticks, some with as many as thirty to forty—embedded ticks,...on private parts and not so private parts. In short order, everyone was buck naked tending to themselves for the intimate parts but then, like baboons grooming each other, picking ticks off each other's backs. It's a good thing that the enemy didn't happen upon us at that moment because we were in exactly the kind of position that I dreamed of catching them in.

Finally the de-bugging process was completed and it was time to move out.

"Saddle up!"

We made the ridge by mid-afternoon and set up a perimeter. Looking north down the opposite side of the ridge we beheld the Song Cu De River. Looking down and slightly to our east, there was a flat, grassy area on the riverbank, standing in distinct contrast to the steep and heavily vegetated terrain on either side. It looked like a huge putting green.

"Check it out. See that grassy spot on the bank of the river? That looks like a likely spot for the dinks to cache weapons brought down the river on sampans. I'm gonna take a patrol down there tomorrow morning and check it out."

The next morning, I called Captain Thomas on the radio to tell him of my intention to take a patrol down to the river to check the grassy area for cache sites.

"Charlie-six, this is Charlie one-zero, over."

"This is Charlie-six, over."

"Roger Charlie six, I'm going to take a patrol down to, I Set: Papa Oscar: Juliet-Tango-Foxtrot-Mike-Lima-Zulu to check for cache sites, over."

There was a pause while the CO's RTO decoded the map grid co-ordinates.

"Charlie One-zero, this is Charlie-six. You better not fuck it up, over."

"Roger, out."

"Fucking ass-hole," I said in my fury as I threw the radio handset to the full extension of its cord.

This was typical of the way Captain Thomas talked to all us lieutenants. This many years later it is tempting to engage in psychological analysis about why he treated us that way, but at the time, it just made me angry. I disliked him intensely and had no respect for him as a commander.[2]

For the patrol, I decided to send Sergeant Stringfellow's squad. He had just taken over the squad from Sergeant Rodriguez and I wanted to check him out. I gave him the mission but left the planning up to him. I would accompany him and observe. He would need to select a route from our current position atop an 868-meter high ridge down to the river's edge. We would leave our rucksacks in the perimeter so it wouldn't be as grueling as the trip up.

It was approximately three kilometers to our destination and "Stringbean" picked a good route, moving with the flow of the land

---

2. Sometime later, when we were in the base camp, I along with the other two lieutenants, confronted Captain Thomas about the way he treated us. He lightened up considerably after that. A few years later, Bill Weatherford and I, now promoted to captain, ran into him, still a captain, at Fort Benning. We were happy enough to see him but he didn't seem too interested in engaging us as equals.

rather than unnecessarily tiring his soldiers by going across the grain of the land. It was about ten o'clock when we reached our destination.

What had looked like a putting green from a distance of three thousand meters and an elevation of eight hundred meters, was, on arrival, a solid, ten-foot-high wall of elephant grass. With our sleeves rolled down to protect our arms from the razor sharp blades of grass, we plunged in. Completely surrounded by grass and unable to see anything else, the only way to navigate was by compass direction. The density of the grass allowed for no movement of air and it was stiflingly hot.

After tromping around for about an hour and finding no sign of human activity, we took a break. I lit up a smoke and was sitting back resting when I felt something tickle my right arm. 'Must be a loose thread', I thought as I scratched my arm. A minute later my arm tickled again and again I absent-mindedly gave it a scratch. A few moments later, an engorged leech appeared, inch-worming its way out from under the cuff of my sleeve onto my hand. I casually touched my cigarette to it, causing it to recoil and regurgitate my now congealed blood onto my hand. I flicked the leech away and wiped my hand on the leg of my pants. Leeches had been a fact of life from day one in the bush. Seldom seen, they would enter through any available opening, to include the lace holes of our boots. Their saliva contained some form of anesthetic, making their bite painless. The only evidence of their visit would be bloody socks and sores at the site of the bite. How they made their exit after feasting remains a mystery to me.

I told "Stringbean" to head on back to the perimeter, taking care to return by a different route than the one that had brought us here. By early afternoon we were back in the platoon perimeter. The other squads had been patrolling along the ridge, finding nothing. Another day, another dry hole.

11. RE-SUPPLY, AN UNEXPECTED GIFT AND GAS

Every three days, we were resupplied with food, ammunition, radio batteries, smoke grenades, medical supplies, and any other equipment that had been expended, damaged, or lost. Additionally, we received mail and a sundry pack containing tobacco products, shaving cream, pens, writing paper, etc. It was a big deal and required that all leaders keep track of what had been used so that the re-supply order would be thorough and submitted on time. Re-supply was organized in the base camp by the company Executive Officer (radio call sign Charlie Five) and delivered by helicopter. Because it was delivered by air, we had to make sure that we were near a fairly open area where the Log Bird (Logistics) could land or, if we were deep in the jungle, we would have to hack out of the jungle a clearing big enough for a helicopter to land. Immediately after re-supply was received we had to burn all cardboard, bury any unused items, and move to a new location given that our location had been advertised to any enemy in the area. All in all, re-supply was kind of a pain in the neck.

At the beginning of February, still high in the mountains, we took re-supply on a barren, rocky ridge.

"Charlie One-zero, this is Charlie Five, over."

"Ahh, this is Charlie One-zero, Alpha, over."

"Roger, One-zero Alpha. Notify your actual that we are zero-five mikes from your loc, over."

"This is Alpha, WILCO, over."

"Five, out."

"Hey, LT", Cameron called to me, "the re-supply bird is five minutes out."

"OK. Call Sergeant Christiansen and tell him to get his squad ready to off-load. You stand by on the radio to confirm the smoke once they are in-bound."

A couple of minutes later, Charlie-five called again, requesting us to pop smoke. Cameron yelled out to Sergeant Christiansen,

"Pop smoke."

In an instant, purple smoke started billowing up from the landing zone. Charlie-five came back on the radio,

"I identify Goofy grape."

"Roger, Goofy grape," Cameron replied.

We could hear the thumping of the rotors and, in another moment, the bird was in view. It vectored in on the smoke and the designated soldier standing in the goal post position. As soon as the skids touched the ground in a torrent of wind and dust, Sergeant Christiansen's squad leaped to their task of off-loading the bird, which was stacked from floor to ceiling with cases of C-rations and other supplies. In addition, two replacements hopped off the bird and were quickly directed away from the flurry of work. A sitting helicopter is a lucrative target, so the immediate task was to empty the bird so that it could take off. That done, Sergeant Rodriguez supervised the sorting of the mountain of supplies into piles for each squad and the platoon command group. Details of soldiers then transported each pile to its destination, where it was further broken down to each individual recipient.

Receiving mail was one of the highlights of resupply, especially love letters from wives and girl-friends. Love letters often came swathed in perfume and across the seal, lip marks and the initials SWAK[3]. For the squad leaders, handing out the mail was almost as much fun as receiving it.

"Tetrault, oh Tetrault, hurry up, this one is too hot to handle" Sergeant Christiansen cooed as he sniffed the letter before handing it to the blushing Tetrault.

In addition to receiving letters from my wife and parents, there was a canvas mail bag with a shoe tag bearing my name. 'What is this?', I wondered.

I opened the bag, to find a quart bottle of Ancient Age bourbon, an apparent expression of Baker's gratitude for giving him the job in the rear. I greatly appreciated the gesture, but what was I going to do with a bottle of whiskey out in the field? I was too fond of the "Water of Life" to get rid of it, but too sensible to want to drink whiskey in the field. My training as an officer didn't cover this situation but, since it wasn't an emergency, I would decide what to do with it later. I stuffed the bottle into my already heavy and over-stuffed rucksack.

While supplies were being distributed, a large pit was being dug. Once distribution was complete and everyone had packed their ammo, rations and other supplies, everything that was left over was thrown into the pit. Cardboard ration boxes, unclaimed rations, shaving cream and other extraneous sundries were all thrown into the pit. Also thrown in were used radio batteries, shorted out first to prevent them being used by Charlie. Once everything was collected in the pit, it was lit on fire and everyone kept a respectful distance, for soon the rejected cans of food and cans of shaving cream would start exploding from the heat. Once everything had burnt down as much as possible, the pit was covered over and we moved on to a new position.

---

3. Sealed With A Kiss.

At this early stage in my career as a soldier, I was still madly in love with the Army and, as in all new love affairs, my lover could do no wrong. I was beginning to wonder about certain things, though. For example, it occurred to me that our method of re-supply was not very subtle. How in the world were we going to catch Charlie by surprise when we boldly announced our presence and location every three days?

***

A couple of nights later, higher up on the same rocky ridge, we were set in a perimeter. Olson, a dour, reliable soldier, reported to his squad leader that he heard movement. Sergeant Stringfellow came to my Command Post position in the perimeter.

"LT, Olson heard movement."

"Where?"

"To our east, down in the draw."

I went with him back to his sector of the perimeter. I couldn't hear anything but Olson insisted that he had. Stringbean asked me, "You want us to light it up with a mad minute?"

"No, that would only give our position away. Go to one hundred percent alert. I'm going to call in arty."

Back at my CP, I called Captain Thomas and advised him of the situation and my intent to call in artillery. I then called the Artillery Forward Observer, who was part of the company command group and whose job was to coordinate our artillery fire support. He was also a lieutenant, affectionately known as Ti-Ti[4] LT because of his diminutive size.

"Redleg Three-one, this is Charlie One Zero, fire mission, over."

"This is Redleg Three one, send it."

---

4. Ti-ti was Vietnamese slang for small. It was derived from the French word 'petite'

"Direction, one, two, zero degrees, distance four, zero, zero meters, target: suspected troops in the open, over."

"Roger, One Zero, on the way, over."

"This is One Zero, standing by."

A couple of minutes later, five rounds came crashing in. They were close but a little too far away from where the noise was heard.

"Redleg Three one, this is Charlie One zero, right one hundred, drop five zero, fire for effect, over."

"Roger, right one hundred, drop five zero. Rounds on the way."

Having made my final adjustment, the artillery battery now fired the full mission and rounds came crashing in for several minutes. In the silence that followed, several soldiers claimed to hear screams. Was it over-active imaginations at work? We'd have to wait until morning to find out. In the meantime, we spent a jittery but uneventful night.

In the morning, I took a patrol down the draw to check out the area where we had called the fire mission. The vegetation was so thick, though, that it was impossible to tell where the rounds had hit and we got no clues to satisfy ourselves or the artillery, who always liked to hear about the results of their work.

Still lugging the as yet unopened bottle of whiskey when we stopped in the afternoon, I decided it was time to do something with it. When I met with the squad leaders to discuss plans for the night and next day, I doled out a thimble full to each and had one myself. Straight whiskey in an immature palate in sweltering heat is anything but a pleasure. This was like a millstone around my neck—on the one hand it was a gift, and a nice one. On the other hand, a bottle of whiskey had no place in the field and, besides that, it was heavy. While we were sitting and talking, all four of us simultaneously sniffed the air.

"What the…Gas, Gas!"

It was tear gas.

As the warning call continued to be called out, everyone was madly digging in their rucksack for that never used item of equipment which was at the very bottom of the bag. In addition, there was great apprehension as to what this meant. Were we about to be attacked? Once masked, everyone stood-to with weapons at the ready. Nothing happened. After a suitable period of remaining masked, someone checked to see if the gas had dissipated[*5], which it had, and the "All clear" was given.

As I re-packed my rucksack, I came to the bottle of Ancient Age.

"Fuck this," I said, as I threw it down the side of the hill where it smashed on some rocks.

---

5. The standard method for checking was to designate a soldier to quickly crack and re-seal his mask and wait to see what happened. If nothing, he would crack it again for a longer period. If still nothing, the 'All clear' would be given.

12. A BATH AND A HOME COOKED MEAL

We were now about three weeks into the mission and had not bathed nor changed our clothes since we deployed on the 8th of January. In the draw below the ridge we were on was one of the many streams that made their way down the mountains to the valley below. We had patrolled the area thoroughly and found no sign of enemy activity so I decided to let the men take turns going down to bathe. There are no "time-outs" in war so even this activity had to be done with care and precaution. A squad at a time went down and half stood guard while the other bathed.

I took my turn with the last squad to go down. Leaving my rucksack in the perimeter, I took just my rifle, a bandolier of ammo and my canteens to fill. It was quite hot and I had worked up a good sweat by the time we reached the stream, even though it was only about a hundred meters down the side of the ridge. This stream was typical of the mountain streams in this little corner of Vietnam. It was about five meters wide with crystal clear, cool water cascading over boulders, rocks. and gravel. Crossing one on the move, we would often dip our bush hats in the water and literally drink a hat-full of water.

Joining the first half of the squad to bathe, I stripped and stepped into the stream, selecting a spot where the water pooled and I could sit down and completely submerge myself.

"Ahhhhh", I thought, as my hot, sticky body slipped beneath the clear, cool water. As the water covered my head, I just sat there for a moment and marveled at how sublime the most mundane of activities can be if you've been without it. My reverie lasted but a few moments before I returned to the reality of the situation and got down to the business of soaping off and getting clean. Back on the bank I shaved, toweled off, and slipped back into my jungle fatigues.

Now that my body was clean, I was sensitive to the touch and smell of things that weren't clean. My fatigues felt very stiff against my skin as if they had been heavily starched. This, of course, was due to the buildup of salt in the material from days and days of sweating. I now became aware of the putrid smell of the towel which I always had draped around my neck, permeated as it was with ammonia and other bodily excretions. Oh well, I'd get used to it in a few minutes.

I went upstream slightly from the bathers and filled my canteens and then took my guard position while the remainder of the squad had their baths. Relieved momentarily from the pressures of being platoon leader and being left with my thoughts, I found myself filled with admiration and affection for the platoon. I knew them all by now, some better than others, some I was closer to than others, but collectively, it, the platoon, was a living organism of which I was a part. The sheer physicality of our lives was a powerful bonding agent. Every day we hoisted these mountainous packs on our backs and climbed up yet another hill. Every day we faced danger together. Every day we shared our food and drink with one another. I felt proud to be part of them.

The last man finished bathing and it was time to get back to reality. By the time we climbed back up to the perimeter, I was sweating profusely and the feel of my fatigues and the smell of my towel had returned to being the norm.

***

A couple of days later, we got a call on the radio from 3d Platoon.

"Ahh, Charlie one zero, this is Charlie three zero, over."

"This is Charlie one-zero, go."

"Roger. We shot a wild boar and have more meat than we can use. Would you like some? Over."

"Roger that. Send me your location and I'll send a patrol over to pick it up, over."

As soon as we got their location I sent a patrol over to get the meat. Meanwhile, the possibility of barbecued pork galvanized the Mexicans. Under the supervision of Sergeant Rodriguez, Garcia, Granado, Salazar, and Sosa built a small fire and got it reduced to coals. They laid their machetes on the coals to create a grill and then they created a chili basting sauce from the jalapenos and salsas that had been sent to them from home.

After the patrol returned with the meat, I took another patrol out to scout the area and left the remainder of the platoon to secure the perimeter and barbecue the meat. By the time I returned, it was ready.

"Here you go, LT, you want a slice?"

"Hell yeah."

Granado carved off a big piece from the haunch and handed it to me.

"All right! Every day's a holiday and every meal's a banquet."

"There it is."

I took a big bite but the spirit of the boar was resisting my advances.

"Damn, this is pretty tough."

Granado laughed and said,

"Con heuvos, LT, you can do it."

I laughed back and renewed my assault, finally managing to tear off a bite size portion. It tasted good, a tribute to the culinary efforts of the cooks, but it was tough as hell, as tough as the life of a wild boar in Vietnam.

All around the perimeter, the guys were enjoying this unexpected treat. Tough it may have been, but no Texas barbecue house could have topped it.

There were no official time outs in this war, but sometimes you had to make your own—cautiously.

Preparing haunch of wild boar. Granado center rear, SGT Christiansen right foreground

13. MONSOON R & R

(REFIT AND RE-TRAIN, NOT REST AND RECREATION)

On the night of 3 February, I received a most unexpected radio message from Captain Thomas.

"At first light, move your element to a Papa Zulu and stand by for extraction. Returning to rear for five-day stand down."

A five-day stand down!? This was a complete surprise. We had fully expected to remain in the field until the third week of February when this mission was scheduled to end. Would we get to go to China Beach for an in-country R & R? Would there be any floor-shows?

I identified on the map a spot within an hour's march that would accommodate four birds for the platoon's extraction. Gathering the squad leaders, I gave orders for the movement to the PZ and the order of flight to the rear. In keeping with the dictum that the leader be the first one in and the last one out, Sergeant Rodriguez would be on the lead bird and I on the last.

By 0900 we were assembled on the PZ in helicopter loads. The platoon was down to twenty-two soldiers and it was easy to fit the whole platoon onto the four Hueys that came to extract us.

The move back to the base camp was uneventful. At the edge of the Camp Crecenz chopper pad, the XO, Lieutenant Jim Cole was waiting to run a rod down the barrel of each weapon to insure that none had a round in the chamber. The first task was the cleaning and inspection of weapons before they could be turned in to the arms room. At the same time, Jim Cole had to orchestrate the turn-in of all ammunition, hand grenades and pyrotechnics for safe keeping until we returned to the field. It was afternoon before these necessary chores were completed and we could see about getting ourselves cleaned up.

Once the company was released to get showered and into clean fatigues, I made my way to the lieutenant's hooch where I found my fatigue uniforms laying on my bunk, cleaned and pressed. Under the bunk was my other pair of boots, all polished and gleaming. This was the work of Hai, the hootch maid for the Lieutenant's hootch whose job was to keep the hootch clean and see to the laundry needs of the lieutenants. For this service, each lieutenant paid her a certain amount per month. Being an egalitarian sort (and a little tight with my money), I rejected the idea of having valet service and indicated early on that I didn't want anyone else doing my chores. So finding my kit all laid out took me aback. If Hai was going to take care of me without any compensation, I would have to re-think my position. I began paying her well for her services. At some level, I began to understand that this was her rice bowl and that her livelihood was more important than my stubborn egalitarianism.

After showering, shaving and getting into a clean set of fatigues, I headed back to the company area where a meal of fried chicken, mashed potatoes, and gravy was being served. There was also a big ice chest full of cold beer and sodas. We felt like kings sitting around in the company street feasting on all we could eat and drink. After dinner, I was sitting around having a smoke with Sergeants Christiansen and Rodriguez, as well as Bishop, Hanesco, Doc Kroze and Garcia when a Korean photographer snapped a picture of us. This photograph returned to each of us later as a 5x7 hand-painted picture bound in a leather frame. This stand-down was turning out alright.

Later on that evening, I made my way to the hut that was euphemistically called the "Officer's Club" and engaged in the standard activity for passing the time and beating the heat—I drank. Sitting at the bar, I came under the scrutiny of the Staff Sergeant who managed the club. He was assigned to Charlie Company but had some medical problem which precluded him from field duty. He was allegedly a Korean War veteran and, according to him, had served in that conflict in the legendary 187th Airborne Regimental Combat Team. Noticing the paratrooper wings on my cap, he chose me to be the recipient of his oft-repeated monologue:

"Well, good evening, Lieutenant, how is everything out in the field?"

"It's OK."

"I sure wish I could be out there with you all. I sure do. You know, in Korea I was with the Rakkasans. You know, the 187th RCT."

"No kiddin'? Yeah, I know all about the 187th. A lot of the senior non-coms in my last unit had served with it."

"That was a STRAC outfit, let me tell you. Westy was the commander, you know. Yes, sir, that was one fine outfit. But I sure wish I could be out there in the field with you all."

"Well, we have plenty of room for you. Feel free to join us any time."

"I'd really like to, Lieutenant, but I've got this medical profile that won't allow me to. But I'd sure like to."

He ran this line one too many times in the presence of Sergeant First Class Randall "The Bear" Rollins, also a Korean War vet and a hard man from the mountains of West Virginia who wouldn't have dreamed of riding a sick slip to get out of the field. A few days later, when we were getting ready to go back to the field, SFC Rollins presented SSG Club Manager with a fully packed rucksack.

"All right, mother fucker, you want to go to the field so bad, here's your fuckin' ruck. Let's go."

Quick as a flash, SSG Club Manager produced the paperwork which excused him from field duty. His absence from the field was no loss to the unit and The Bear's confrontation with him had the useful effect of shutting him up about how much he wanted to be in the field.

Several drinks later, weary of SSG Club Manager's bullshit, I made my way back to the Lieutenant's Hootch. After many days of sleep deprivation and the effects of the alcohol, I was sound asleep in no time. Sometime later, I heard a whispering through the fog of deep sleep.

"LT! Hey, LT."

"Hmmmm?"

"LT."

"Hmmmm. What?"

I groggily opened my eyes and there was Bugge and Warner crouching by the side of my bunk. Struggling back into wakefulness, I mumbled:

"Hey. What's up?"

"Nothing. We just thought we'd come by and see how you were doing."

Warner was a wild cowboy type from Colorado. He had been in the Army and in Vietnam for a few years and had apparently risen as high as Staff Sergeant before being busted down to Private for reasons unknown to me. He was a good soldier who frequently walked point. He seemed to relish the war and its connection to his name. He wore a bracelet that said "War" and had a "War" tattoo. For all of that, he was a happy go lucky sort and not at all vicious.

Bugge was a freckled, carrot-headed hippy from San Rafael, California. He was easy going and, in spite of his peace and love past, a good soldier.

Wearing only a T-shirt and drawers, I swung around and sat on the side of my bunk, reaching immediately for my cigarettes.

"Want a smoke?"

Both accepted the offer and we all lit up.

"So what's going on?"

"Oh, nothing. We had nothing to do so we stopped by to shoot the shit."

"What did you guys do tonight?"

"There was a floor show in the EM club. Filipino group. You know, the usual."

"What did you do, LT?"

"Nothing much, just a few drinks at the O-club and I hit the rack."

"So how are you liking the platoon?"

"It's great. I didn't know what to expect after all the shit I'd seen in the papers. But no, I like it a lot."

"There it is. I'd like to get those fuckin' reporters over here and get them out in the bush."

"Well fuck it, it don't mean nothin'."

"There it is."

"Well, I guess we ought to be going. Thanks for the smoke."

"All right. Thanks for stopping by. See you in the morning."

I finished my cigarette and lay back on my bunk with a warm feeling that all was well in my world.

The next morning, I went to the mess hall and had the first real breakfast that I'd eaten since January 8th. Real coffee, scrambled eggs, bacon, and toast. Breakfast had always been my favorite meal of the day and this one really hit the spot.

After breakfast, Lieutenant Tom Mylan and I wandered back to the hootch, only to discover that Major Lenhart, the Battalion XO, was there doing an inspection. Tom and I joined the little knot of lieutenants already there enduring his tirade. This was the same Major that I first encountered on Christmas-eve when the hand grenade was thrown under the hootch.

"Get these beds properly made. What do you think this is, a fraternity house? And get all that crap off those shelves."

On the wall above each bunk was a small wooden shelf. Glancing at the shelf above my bunk, I noticed a few of my leftover cans of C-ration food, to be eaten as snacks or taken when we went back to the field. After the Bruiser, as he was known, stormed out, we commenced to grumbling.

"What a prick! What is this, basic training?"

"What the hell is the point of having a shelf if you can't put anything on it?"

"He's probably got nothing better to do than fuck with people."

"Well fuck him."

"Yeah, let's give him an Army Hymn."

"Him, Him, Fuck Him," we chimed in unison.

Next I strolled down to the latrine for my morning constitutional. The latrine was a five-holer bench raised over halved 55-gallon drums. From this perch there was a majestic view of rice paddies and the city of Da Nang in the far distance. There was no toilet paper but there were plenty of copies of the 'Stars and Stripes' newspaper. One minute it would be a source of news, the next, toilet paper.

Some time that morning, I managed to get patched through by telephone to my friend John Beasley, who was assigned to a liaison unit in Hoi An, about 30 miles south of Da Nang. The connection was scratchy and faint, but sufficient.

"196th Liaison, Lieutenant Beasley speaking, sir."

"John, it's Buzz."

"Buzz! How you doing? It's good to hear from you, man."

"I'm doing fine. Hey, listen, we're on a five-day stand down. Why don't you try and come on up here for a visit?"

"I could probably swing that. Things are pretty quiet here and I have my own wheels. Let me see what I can do."

"Great. So hopefully I'll be seeing you in the next couple of days. I'm in One-Four-Six at Camp Crecenz."

"Yeah. I'll try to get word to you when I'm coming."

Later, in the company orderly room, I was talking to Lieutenant Jim Cole from whom I had taken over the platoon and who was now the company Executive Officer.

"So how is life in the rear?"

"Well, I'm glad to be out of the field, but sometimes that seems better than being in the rear."

"How so?"

"The drugs, for one thing."

"Is there a drug problem back here?"

"Come on, let me show you something."

He got up from his desk, put on his cap and crossed the company street to the hootch where the cooks, derisively known as 'spoons', lived.

"This is the spoon hootch. Take a look underneath."

All the hootches were raised on blocks, so I bent down and looked at the twelve-inch space under the floor. It was littered with little plastic vials that resembled the cases that people kept their contact lenses in.

"What are those?"

"Those are heroin vials. They shoot up in the hootch and drop the vials through a hole in the floor. That's only part of what we put up with in the rear."

"Fuck me. What else?"

"The other thing is all the Black Power shit, dapping and refusing to salute officers. It's a pain in the ass. The other day Schexnayder was dapping with some other dude when Captain Thomas was walking by and they ignored him. He lit them up like a Christmas tree. But he's black and can get away with it. It's not so easy if you're a white lieutenant."

"Wow, that sucks."

I had already gotten an inkling of this tension before we went to the field and had made it a rule of thumb to try and avoid those confrontations by not knowingly blundering into one. As an officer, I was duty bound to uphold discipline and call an infraction, but if there was no infraction, there was no confrontation.

***

The next day, the three platoon leaders, Lieutenant Tom Mylan, Captain Crunch and myself, were hanging around the company headquarters hut without much to do. Platoons are supposed to be led by lieutenants and Crunch was a captain but was so eager to be in combat that he begged to be allowed to lead a platoon. A Texan, he was all enthusiasm and no common sense, which was how he came to have the nickname Captain Crunch.

Shortly, the CO, Captain Thomas appeared and said, "Let's take a ride into Da Nang."

The closest I had been to Da Nang was the village we called Dogpatch, I had never been to Da Nang and I really didn't have anything else to do. Also, it would have been bad form to snub the CO by rejecting his suggestion.

"Yes, sir."

"We'll leave in ten minutes."

After taking care of a couple of quick errands, I reported back to the company HQ. Captain Thomas had gotten his jeep and was going to drive. Tom Mylan and I, being the junior members, crammed into the back seat. The only person missing was Captain Crunch. Within moments he appeared coming from the direction of the arms room and I noticed that he was now armed with a .45 caliber pistol. The city of Da Nang was under government control and it wasn't necessary to go there armed. I wasn't particularly surprised at this precaution because, after all, it was Captain Crunch.

Off we went, down the road through the familiar sights, sounds, and, most of all, smells of Dogpatch. It was a squalid little village which stank of rotting fish, rotting vegetables and general dirtiness, mingled with the exotic smells of unfamiliar food being cooked. The open fronted houses were situated right at the edge of the road where there was a constant stream of foot traffic—men dressed in shorts and pith helmets carrying loads on both ends of long poles which were balanced on their shoulder. The frequent passage of military vehicles stirred up clouds of dust, which settled on everything and everybody.

It was a dreary, overcast day and, as we entered the outskirts of Da Nang proper, I was struck at how equally dreary the city looked.

'Where was the exotic Asian architecture?' I thought. 'Where was the stately French colonial architecture?'

All I saw was drab, nondescript buildings surrounded by drab, nondescript streets and alleys. Maybe it would get better as we got deeper into the heart of the city.

All of a sudden, a blur appeared at my right and Captain Crunch, sitting in the 'shotgun' seat right in front of me, gave out a yelp. I took me a moment to realize that the blur had been the hand of a kid and that the yelp from Crunch had been his response to his watch being snatched off his wrist.

Captain Thomas slammed on the brakes as the rest of us rose as one to pile out of the jeep. Captain Crunch was bellowing like a maddened bull as two little kids disappeared down an alley.

"God damn you, you sonofabitchin' blue-balled bastards! Give me my watch back."

Continuing his bellowing, he ran down the alley, pulling out the .45 as he ran.

"God damn it. I want my watch back. I better get my fuckin' watch back or it's going to be hell to pay."

The rest of us followed him into the alley, which was eerily silent and devoid of any human activity. I was thinking how surreal the situation was. We'd left the jeep unattended out in the street and were in an empty alley with big buildings all around and only one of us armed and him seemingly capable of doing something really stupid, so enraged was he.

I was thinking, 'I'd sure hate to get killed in a back alley in Da Nang over Captain Crunch's crumby watch.'

Then, miraculously, an old papa san emerged from one of the buildings and brought the watch back to Captain Crunch. He must have been thinking along the same lines as me, that it wasn't worth dying over a watch. I don't even remember the rest of our trip into Da Nang. Whatever happened, it wasn't nearly as interesting or engaging as the great watch snatch.

During morning formation, the First Sergeant had a surprise announcement:

"Alright gentlemen, listen up! After formation, we are going to have a mandatory piss test. Every swingin' dick of you is going to give a urine sample. And when I say everybody, I mean everybody — the CO, me, all officers and NCO's. Is that clear?"

"Yes, First Sergeant!" came the shouted response.

"We WILL extract urine from each one of you and your officers are going to supervise. If you don't have to go, you better start drinking water because nobody is going anywhere until every one of you has given it up. Is THAT clear?"

"Yes, First Sergeant!"

This was part of the Army's effort to stem the growing use of drugs by soldiers. Officers were not exempt. We led the way and then had to personally supervise the urine collection for each member of our respective platoons. This was an unpleasant task but one I didn't particularly object to as I was fully supportive of any measures that would discourage drug use. So far, I had had no indication of drug use in the platoon and I wanted to keep it that way.

After the collection was finished, I received a surprise visit from my friend John Beasley who was stationed at Hoi An, about thirty kilometers south of Da Nang on the coast.

"Beas! Hey man, It's good to see you. How did you get up here?"

"I've got my own jeep."

"Oh, yeah, I forgot."

"What do you want to do?"

"Since you've got wheels, why don't we go down to the air base and have a bite and shoot the shit?"

"OK, that sounds good."

For the second day in a row I was making the trip down the road from Camp Crecenz and through Dogpatch. Instead of turning south toward Da Nang, we kept straight on to the big air base, which I had seen only once before when we air assaulted into it. Once through the gate, it was as if we had miraculously been transported onto a stateside military base. Instead of wooden huts, there were permanent buildings, three-story concrete dormitories with air-conditioning units in the windows, clubs, restaurants, everything.

"Wow, look at this. The Air Force's got it rough, don't they."

"Look at this shit! Fuckin' air conditioning and everything. I don't even think of the Air Force as a military organization. They're just fuckin' civilians wearing blue suits."

"There's a restaurant. Why don't we go there and get some lunch?"

Once inside and seated, we both ordered shrimp fried rice and a bottle of champagne. I was eager to find out about John's new assignment.

"So, how's your job? Are you an advisor to the ARVN*?"

"Not exactly...I'm assigned to the 196th and supposedly we are liaison to the ARVN. Some of the work is interesting but actually, there's not much to do."

"Hey, why don't you get yourself transferred up here? You could get another platoon. We have a shortage of lieutenants. Shit! That would be fuckin' great."

"I don't know. I'll think about it. It's kind of nice having a break from humping the bush, but I am kind of bored."

"Well think about it."

Mellowed by the champagne, we spent the rest of the meal reminiscing about our time together in the 82d Airborne Division, laughing about the many funny things that had happened and feeling like old soldiers, in spite of the fact that neither of us had yet been in the Army for two years.

After lunch, we drove back to Camp Cresencz, where Beas dropped me off at the gate.

"Take care, Buzz, I'll stay in touch."

"Yeah, thanks again for coming up. Give some serious thought to transferring up here."

"OK, I will."

***

"COMPANY,…TEN…HUT! REPORT!

"First Platoon all present or accounted for, First Sergeant."

"Second Platoon all present or accounted for, First Sergeant."

"Third Platoon all present or accounted for, First Sergeant."

I was standing at the back of the company formation as the First Sergeant took the morning report and issued instructions for the day.

"AT EASE AND LISTEN UP! We're heading back to the bush tomorrow at zero-seven-hundred. We will draw fatigues immediately after formation, company headquarters element and First Platoon at zero-eight-thirty, Second Platoon at zero-nine-hundred and Third Platoon at zero-nine-thirty. Platoon Sergeants, turn in your requests for rations, ammo and batteries by ten-hundred-hours. Platoon leaders will be meeting with the old man at ten-hundred and any orders from the officers will be after that. COMPANY…TEN…HUT! FALL OUT!"

I joined my platoon at the supply room where we filed past the distribution window and received a set of fatigues from Specialist Baker, formerly of First Platoon and now the Supply Clerk. In the rear, we wore our own fatigues with our name and all patches and insignia sewn on. When we went to the field, we drew from the supply room a set of "field fatigues" which were generally devoid of any markings or insignia.

"Hey LT, how you doin'? How did you like that Ancient Age I sent you?"

"That was great stuff, Baker, except that the old man found out about it and reamed me out."

"Fuck it, sir, it don't mean nothin'. I mean, what are they goin' to do to you, send you to Vietnam? What size do you take?"

"Give me a large, thanks."

I took the fatigues and headed back to the hooch to deposit them there. When I got there I checked the size, just to be sure. The jacket was a large but the pants were a medium.

'Aw fuck. Maybe they'll fit. Let me try them on.'

I slipped them on. They weren't super baggy like a large but they seemed to fit. In fact, they seemed kind of stylish. It's amazing what can pass for style when everyone wears the same thing. I didn't have a mirror but from what I could see and feel, I began to fancy myself a regular Beau Brummel in the fashion world of olive green fatigues.

'Yeah, these will be fine.'

At 1000, I reported to the orderly room for Captain Thomas' instructions. Before talking to us collectively, he addressed me.

"Looks like you're going to be needing a new RTO."

"Why's that, sir?"

"Your boy Cameron pissed hot for marijuana and has had his security clearance pulled. He'll stay with your platoon but he can't be the platoon RTO. I'm sending you Gadzinski from the CP element as his replacement."

"Is he any good?"

"Of course. I'm sending him from the CP element, aren't I?"

I was suspicious. Captain Thomas and I didn't get along very well and, over and above that, it was rare that a higher headquarters would send someone who was good down to a lower level. *

After dropping this bomb-shell on me, Captain Thomas commenced to address all the platoon leaders.

"We're going to be working the low ground for the remainder of the mission. We're looking for routes of infiltration from the mountains into the villages and we're looking for weapons caches, especially along the blue lines. We'll be going into a company-size LZ here. First Platoon will work this area here, second here, and third here. Any questions?"

The low ground consisted of foothills at the base of the mountains and broad, flat valleys where rice had been cultivated before this area had been cleared of its inhabitants. Flowing down out of the mountains and heading east toward the coast were numerous large streams.

I knew from past experience that this was as much as we were going to get in the way of an operations order and to ask questions would suggest criticism. If I had any questions, I would find out the answers on my own. After collecting what additional information I needed, I briefed Sergeant Rodriguez and the squad leaders on the platoon mission and their specific part of it. The rest of the day was spent in preparation: drawing equipment and supplies, packing and even making time to go to mass.

I had mixed feelings about going back to the field. Part of me was eager to resume the hunt for Charlie and also get away from the boredom and hassles of the rear. But another part of me was not looking forward to humping that heavy rucksack, its straps digging mercilessly into my shoulders. But, after all, this was what I signed up for.

*In the event, Gadzinski turned out to be a superb RTO and we got along very well into the bargain. Beyond the bonds of comradeship and ordinary soldier chatter, he was someone I could talk to at a deeper level than I was able to with most of the guys. When time allowed, we would engage in what were, for our age, deep philosophical discussions about the war, life, etc.

1st Platoon, cleaned up and ready for inspection

SGT Russell "Chris" Christiansen and SGT Ricardo "Rod" Rodriguez

1st Platoon just in from the field

SP4 John Elzey, SP4 Jesse Brown, SP4 Danny Bishop, SP4 Ruben Bugge

14. SPIDER-MAN

The return to the field was uneventful. After landing and getting the platoon in a temporary perimeter, I gathered the squad leaders and platoon sergeant to give them final instructions. I took out my map and squatted down to put it on the ground and orient it to the actual terrain.

Rrrrriiiiiiippppppp...

My pants ripped along the seam from the waist-band all the way to the crotch.

"Oh LT," Sergeant Rodriguez laughed, "your ass is out."

And indeed it was. Since I didn't wear drawers, my ass was well and truly out, giving literal meaning to this common GI expression. So much for my stylish, size medium pants. I wasn't going to be the Beau Brummel of the Bush in this condition. There was nothing to be done for it. I would just have to be careful where I sat down.

We moved off the LZ for two to three kilometers and set up a patrol base on top of a knoll in the foothills at the base of the mountains. Unlike the mountains, the vegetation was sparse, consisting mostly of knee-high grass and scrub brush. The upside of this was unobstructed field of vision, the downside being exposure to the direct sun, making it very hot.

As usual, I sent the squads out to patrol the area in cloverleaf pattern. Very shortly after sending them out I received a call that a spider hole had been discovered. Spider holes were very narrow holes in which the VC could hide or pop out of to shoot at us from and disappear into again. The hole was barely outside our patrol base so it took no time for me to make my way to the scene. The hole was barely discernable in the tall grass and scrub. The immediate precaution was to throw a grenade into the hole, which went off with a muffled crump.

"Where's Houchin? Get him over here."

Houchin was from Oklahoma, a rustic if there ever was one. In addition to having a strong Oklahoma twang he sounded like he had a mouthful of spit when he talked so that his words came out sloshing and gurgling. He had a favorite piece of soldierly wisdom that he liked to dispense to everyone. Holding up his rifle, he would solemnly pronounce, "Shhirr, you take care of it and it will take care of you."

Houchin was also the platoon tunnel rat. He would fearlessly go down holes without giving it a second thought.

We carried one .45 caliber pistol in the platoon for situations like this. Sergeant Stringfellow was the custodian of the .45.

"Tell Stringbean to get over here."

Meanwhile, Houchin showed up and began stripping to the waist.

"Be careful Houchin. Tell us what you find."

"Yeshh shirr."

Armed with a flashlight and the .45, he wormed his way down the hole, which was barely big enough for him to fit in.

"It looksh like itsh connected to other tunnels."

Pretty soon, we couldn't hear him any more and we waited anxiously for him to reappear.

After a while he came back, inching his way feet first. Once his feet poked out of the entrance we were able to grab his feet and haul him out.

"So what's down there?

"Well, there weren't any dinks, but there are five interconnecting tunnels going off in all directions."

Everyone then started scouring the area looking for other spider hole entrances. I threw a smoke grenade down the hole and covered the entrance with a poncho, hoping the smoke would escape from other entrance holes. We were taught this technique in training but we looked in vain for exiting smoke. By searching we found a couple more holes and were able to estimate that this tunnel complex covered an area of about twenty meters square.

I didn't consider it safe to stay in our present location for the night so we called in our find to the Company and then moved on to a new location.

15. MAIL

By now I was getting mail regularly, primarily from my wife Sue and from my parents. We had discovered on my leave prior to coming to Vietnam that she was pregnant, a discovery that gave her a definite boost from her despondency over my going overseas. Her letters, usually scented with Estee Lauder perfume, told of the progress of her pregnancy and of the trials of being back home in New Orleans with her mother. I was also getting periodic "care packages" from her, most recently a box of pralines, which I shared with the rest of the platoon.

Within the bounds of decency, the guys would share their news from home and maybe even a sniff of the envelope.

"Hey sir, who'd you get a letter from?" asked Sergeant Willard one day.

"It's from my wife."

"Damn, sir, you married?"

"Yeah, and she's pregnant."

"You got a picture of her?"

I fished my wallet out of my back pocket and showed him a picture of her.

"Awww, how'd you rate that, LT?"

"Just lucky, I guess. Here, have a whiff," I said as I waved the envelope under his nose.

"Hey, did you hear that Sosa's wife had a baby girl?"

"No kidding. Hey Sosa, congratulations, man."

PFC Eddie Sosa, Dallas, Texas, was in the first batch of replacements to arrive after me and now he was the beaming, proud father of a little girl.

My parents were also going through big changes. My dad, at age sixty-two had pretty much come to the end of the line with his phonograph record store, falling victim to the rise of the big chains which could easily undersell him, and was facing an uncertain future in early retirement. My mother, also sixty-two, was moving from her job at Herrick Hospital in Berkeley into private practice in the office of an orthopedic surgeon who wanted her services.

It was kind of weird to juggle and find balance in my head between all these activities going on at home and the reality and demands of my daily existence. It was also a challenge to decide what to recount in my letters home. I was careful in my letters to my wife not to mention anything that sounded remotely dangerous whereas I was more open with my letters to my parents.

Mail came every three days with re-supply and was a very big deal for everyone.

16. MA

I moved the platoon a couple of kilometers west, out of the foothills and into the heavily vegetated base of the mountains. This offered cover and concealment and also got us out of the direct rays of the sun. The weather could still be cool high in the mountains, when it rained or was foggy, but the arc of the weather was definitely heading in the direction of the hot season.

I selected a spot about fifty meters north of a major east-west trail coming out of the mountains and heading in the direction of Da Nang. After patrolling the immediate environs for signs of the enemy, we set up for the night. I placed a mechanical ambush (MA) on the trail. (An MA was the U.S. Army's version of a booby-trap, a trip-wire activated claymore mine, which blasted 700 steel balls at whatever was in its way). Since there was no sign of enemy activity, I put the platoon on one-third alert.

As darkness fell, I cooked my standard evening meal, ramen noodles topped with a can of 'Beef with Spiced Sauce'. Cooking was done on what was known as a Ranger Stove, which was nothing more than a ventilated C-Ration can fueled by a Trioxine heat tablet. After dinner, Sgt. Rodriguez, Granado, and a couple of others gathered around to talk quietly. Two of the favorite topics were food and when the brigade was going to stand down. We knew it was going to happen but when was the big question. Jeffries had the latest rumor.

"Last time we were in the rear, I talked to a guy who was coming back from Brigade headquarters and the CG came by and gave him a ride. The CG told him that we were going to stand down in March."

"Wow! March? No kiddin'?"

"Oh, bull shit. You think the CG is going to tell Joe Shit the Rag Man something like that? Come on now."

And so it went. Outwardly, I went along with the notion that I couldn't wait for stand down, but secretly, in my heart of hearts, I didn't want it to end. I had been living for this for four years, delaying gratification in order to plod through college and get a Regular Army commission. The thought of having it pulled out from under me was too much to bear. But, of course, I couldn't really say that.

"Hey, let's make a peach cobbler."

"I've got a can of pound cake."

"So do I."

"I've got a can of peaches."

"Me, too."

Everyone scurried back to their positions to gather the ingredients. First, the pound cake was broken up and placed into a canteen cup. Next the peaches were added, and finally some coffee creamer for texture. Stir and serve. We each sat there with our little plastic C-Ration spoons as the cup was passed around, taking a bite when it was our turn and then passing it on. Communal living at its best.

"Oh man, this is fuckin' great."

"It don't get any better than this."

"You know that's right."

"Fuckin' A."

After checking the perimeter, I crawled under my hootch to get some sleep. Fully clothed, boot laces loosened but boots on, rifle by my side. Sleep, under the best of circumstances, was never sound. The mind always stayed half awake, listening to every sound, alerting to anything out of the ordinary, much, I suppose, like a mother listening to her infant child. But tonight, even that level of sleep was impossible.

As soon as I lay down, the mosquitoes swarmed around my face. First I tried to ignore them, then I tried waving them away. Finally, in desperation, I pulled my nylon blanket up over my face. On the other side of the blanket, I could hear hundreds of mosquitoes angrily screaming for my blood. But now a new problem arose — I was suffocating. Having a blanket over my face in this hot, humid air gave the sensation of not being able to breathe. But I was breathing and I knew I was breathing so I tried to endure it, but after a few moments I couldn't stand it and I threw the blanket off my face, only to be assaulted by hundreds, maybe thousands of hungry mosquitoes. I went through this routine for about an hour before I gave it up as hopeless. Returning to the center of the perimeter, I discovered Sgt. Rodriguez and some of the others already there, talking in whispers.

"Fuckin' mosquitoes were eating me alive."

"Were!? They still are."

"Yeah but at least we're not trying to sleep."

"Ay Chingao. No question about who's winnin' this pinche war."

"There it is."

"Well, since we're up, we might as well have a party. How about a café mocha?"

"Can't dance. Fire up the water and I'll get what I've got."

Everyone crept off to their positions to collect their share of the ingredients to this field delicacy. Into a canteen cup of boiling water was stirred a packet of cocoa mix, two to three packets each of coffee, creamer and sugar. Once thoroughly mixed, the cup was passed from person to person until it was almost gone. Sgt Rodriguez held the cup out to me.

"Mate lo, LT."

"Naw. Go ahead."

Sgt. Rodriguez drained the cup as we all sat around inwardly smacking our lips and wishing there was more. But it was good and the caffeine helped us to not miss the sleep which the mosquitoes denied us.

KA BLAM. The sudden noise shattered the night. Quickly orienting on the direction of the explosion it was immediately clear that our MA had been blown. Did they know we were here? Were we about to be assaulted? All these thoughts were running through my mind along with the knowledge that I would have to anticipate the answers and make decisions.

"Get everyone on one hundred percent. Don't shoot unless we get assaulted. We don't want to give away our position."

SGT Rodriguez scuttled away to the squads while I called the company commander and reported in hushed tones what was going on. Very quickly, silence descended on the forest as if nothing had happened. SGT Rodriguez returned.

"LT, you want to send a patrol down to the trail and check it out?"

"No. I wish we had a manned ambush there but, since we don't, I don't want to send guys stumbling around in the dark. We'll check it out in the morning."

"Roger."

We stayed on alert for about an hour, with absolutely nothing happening. I put the perimeter at fifty per cent alert. Between the mosquitoes and the MA going off I doubt there was much sleep in any case.

In the morning after stand-to, I accompanied a squad down to where the MA had been placed, eager to see what damage we had done to the enemy. Nothing. The VC and NVA were famous for removing their wounded and killed but surely there would have been blood or some other clues. But all we were left with was one more story of the phantoms who were just around the corner.

Coming back into the perimeter I heard mild grumbling.

"Ah come on doc, these things give me the shits."

"Maybe so, but that's better than malaria."

"It must be Monday. Doc Kroze is handing out the horse pills."

For the prevention of malaria, we took a small pill every day but once a week, on Mondays, we had to take a gigantic pill, which had a laxative effect. The doc would make sure that each man took the pill

"Morning, LT. Here's your horse pill."

"Morning, Doc. Give me a lomotil too. I have the shits with or without the horse pill."

"No sweat."

A round face, unruly mop of hair and a perpetual smile, Doc Kroze was from a little town near Everett, Washington. For religious reasons, he was a conscientious objector and would not carry a weapon. He was a good medic and a nice guy and everyone respected him.

While my coffee heated over my tiny "ranger stove", I savored my breakfast of fruit cocktail, lingering over the last drops of syrup as I drained the can into my mouth. Next came the bitter warmth of C-ration coffee drunk from an empty C-ration tin, using the bent lid as a handle. Gadzinsky called to me from nearby.

"LT! It's Six for you."

Taking the hand-set, I responded.

"Roger Six, this is One-zero."

"Six here. Change of plans. Move your element to a Lima Zulu. We are going to do Eagle Flights."

Quickly consulting my map, I identified a tentative LZ, saying that I would confirm the location when we got there.

Eagle Flights consisted of flying a unit to an area, doing a quick sweep, then being picked up and flown to another location to do the same. The idea was to keep the enemy off guard and potentially catch him unawares. This was to be my first Eagle Flight operation so I was pretty excited about the prospect of it being successful. Further information from the CO indicated that we were going to Elephant Valley to engage two companies of VC which were reportedly there.

Hurriedly gathering the squad leaders together, I passed on what little information that I had, saying that I would give specific instructions once we got on the ground.

"Alright, saddle up. We're going to un-ass this AO."

17. EAGLE FLIGHTS

I was excited at the prospect of locking horns with the VC. I was also excited about doing Eagle Flights. It was a tactic developed in the 101st Airborne Division in 1966 as a way of being everywhere at once, thereby heightening the possibility of catching Charlie off guard or, at a minimum, keeping him off balance. Since speed of movement on the ground was essential, we were to leave our rucksacks on the helicopters, taking with us only water and ammunition. Our rucks were to catch up to us later in the day. No soldier likes to be separated from his gear but we had to trust that everything would work out as planned.

As the first bird came in, I, along with the other five passengers, ran out to get aboard. With gestures and shouting over the noise of the engine, the crew chief instructed us to pile our rucksacks in the center of the cargo area where he lashed them down with a cargo strap. We took our places, three on each side, sitting on the floor with our legs dangling out. As soon as the crew chief gave the pilot thumbs up, we were airborne and heading east toward the coast, destination Elephant Valley. The exotic name and the fact that we had never been this far east before added to the sense of excitement.

As we made our approach to the LZ, all I could see was a sea of elephant grass, causing me to conclude that the exotic name had nothing to do with the presence of elephants, but with the razor-sharp grass that grew to great heights. From the air, I could see a low hill mass, which was our left boundary, separating us from the other platoons to the north.

After exiting the lead bird, I crouched in the tall grass waiting for the rest of the birds to arrive. As the next one came in, I noted Sergeant "Rock" Mixon riding outside the aircraft balanced on the skid, ready to leap off as soon as it touched ground.

'Yeah, this guy's a stud dog,' I thought.

Sergeant Mixon, of Dallas, Texas, was a recent replacement, having been transferred to me from the brigade Ranger company when it was de-activated. He was a graduate of the Army's Ranger and Parachute schools, making him a member of the Army's infantry elite. I was absolutely delighted to have gotten him assigned to my platoon. In my youthful naiveté, I had a tendency to be dazzled by appearances. I was to learn from experience that all that glitters is not gold.

Once the platoon was all on the ground, I quickly gave orders to the squad leaders.

"The terrain here is wide open and flat, so we'll sweep it on line. First Squad on the left with the hill as your left boundary, Second Squad in the center, Third Squad on the right. Keep everyone spread out, ten meters apart. I'll be behind Second Squad in the Center, Rod, you follow Third Squad on the right. Since there's no real objective, I'll let you know on the radio or by hand and arm signals when to stop or change course. Any questions?"

There being none, I moved with Second Squad as they spread to the right of First Squad. Once everyone was in position, I gave the 'forward' signal with my arm and we began to move forward at a walking pace. Everyone looked intently to their front, with an occasional glance to the left and right to make sure that they were staying on line, not getting ahead or falling behind.

All of a sudden, there was an explosion of small arms fire from the other side of the hill to our north. Assuming that the platoon on that side of the hill had made contact, my immediate inclination was to move to the top of the hill, seizing the key terrain and being prepared to support by fire or maneuver as the situation warranted. Before giving the platoon new orders, I called the CO.

"Charlie-six, this is Charlie one-zero, we have contact to our November, I'm going to move to the high ground, over."

"Negative, negative," came the reply, "that's the one-three element conducting a recon by fire."

"Roger, out."

I was pissed.

"Those idiots. You'd think they would have let us know if they were going to do a recon by fire. They might have lit us up if we had moved up on the hill."

It didn't take long to finish sweeping the area and concluding that no one or nothing had visited here any time recently. The platoons to the north of the hill mass came up just as empty.

We reassembled at the PZ and awaited pickup and transportation to our next destination. Once airborne, we headed due west toward the mountains until there was nothing below but jungle and mountainous terrain. Off in the distance, I saw a tiny puff of yellow smoke emerging from the dense canopy, marking our LZ. Someone had obviously gone in ahead to mark it for the pilots. As we got closer, I could see that it was a one-ship LZ, meaning that there was a hole cut in the jungle just big enough for one bird to descend into—sort of like going down a chimney. Ever the optimist and assuming that everyone was competent at their jobs, it never occurred to me that the slightest error by the pilot could cause a rotor to clip a tree and down we would go like a ton of bricks.

Once the whole platoon was on the ground and in a perimeter around the LZ, I gave orders to the squad leaders to conduct squad size patrols in our usual cloverleaf pattern, using compass bearings and the clock method to identify zones. Because of the thickness of the vegetation, it would be necessary to move in single file.

"Due north is twelve o'clock. Chris, you take from ten to two, Stringbean, take from six to ten, Rock, you take from two to six. Don't go out any further than a klick. We need to be back here and ready to move in no more than two hours. Stringbean, I'll go with you. Rod, you move with Rock."

Again, there were no questions and each squad moved out on their designated leaf of the clover. And, again, there was absolutely no sign of Charlie. So much for these exciting Eagle Flights.

Our next destination was at the base of the mountains where the jungle gave way to scrub brush and tall grass. It was afternoon by now and this would be our last mission of the day. As we approached the LZ, we were preceded by Cobra gunships, which started shooting rockets and firing machine guns all around the LZ. The rockets started fires in the grass, which began quickly spreading. As the Hueys came in one by one, dropping us off, the crew chief kicked out our still stowed rucksacks into piles on the ground.

As always, I was on the lead aircraft. Surveying the scene that was unfolding, I saw the real potential for us to be engulfed in flames if we didn't move quick. Glancing around, I saw that there was a small knoll about fifty meters to the east.

"Grab a ruck, any ruck, and head to the high ground over there!" I shouted, indicating the hill we were to move to.

Everyone, as they arrived, saw the danger and moved with a purpose. Once I was sure that everyone was in motion, I moved to the ruck pile and grabbed one. I had the acute misfortune of grabbing the bag belonging to a six-foot-four machine gunner who was also a consummate chow hound. In addition to his water and machine gun ammo, he must have had an entire case of C-rations in there. I could barely hoist it onto my shoulders and, when I did, it sank down to my butt because the straps were extended to accommodate his gigantic frame. I literally staggered the fifty meters over to our hill and up it. Once at the top we established a ragged perimeter around the top, whereupon we turned our attention to the fire, which was heading in our direction.

"Everyone grab an E-tool and get down there and start fighting the fire!"

This was clearly an 'All hands dance' situation and every one of us manned the fire line, cutting a fire break between us and the fire as well as shoveling dirt and stomping it out as it got within range.

Finally, by late afternoon, we had it under control and were safe on our hilltop. Now it was time to get back to the business of the war. Rucksacks were reunited with their owners, squad sectors were identified and properly manned, and radio watches were set for the night. We were back in business. Gadzinski and I were together in the platoon CP in the center of the perimeter. As darkness descended, Gadzinski and I brewed up a cup of coffee and discussed the events of the day.

"Man, LT, that was pretty messed up the way the gunships started everything on fire. We could have been in a world of hurt."

"Yeah, you know how the aviators are, they probably thought it was funny. Also, it's really gotten me to wondering about the way we operate. I mean, like the way we do our Combat Assaults, we might as well have a brass band out to announce our arrival. It's no wonder that we can never find Charlie."

"It doesn't make a whole lot of sense to me either."

"Well, I think I'm going to get some sleep. I'll be right here. Make sure the radio watch knows where I am and wakes me up if anything happens."

"Roger that, LT."

My next sensation was a feeling of warmth and an awareness of light. Then I became fully conscious. It was broad daylight and the sun was beating down on me. Quickly looking around I realized that I was the only one in the platoon who was awake and I had just woken up.

"Jesus Christ! Fuck! Gadzinski, wake up!" I blurted as I shook Gadzinski awake. I then ran around the perimeter waking everyone up. It was immediately obvious what had happened—in our exhaustion after fighting the fire everyone, myself included, had succumbed to sleep. There was no blame to be assigned except to myself, since the platoon was my responsibility and I had not foreseen the need for extraordinary measures to make sure that the watch was kept. We were very lucky that there were no unwelcome visitors during the night.

Later in the morning, Gadzinski and I were sitting on the side of the hill waiting for the birds to pick us up for our next adventure. It was late morning and already so hot that I could hardly breathe.

"Hey, Gadzinski. I've been reading this book about the Indian Wars. It wasn't like what you see in the movies. The Indians were way better horsemen than the Army and they knew the terrain like the backs of their hands. Most of the time, the Army just rode around looking for Indians but never finding them. Sound familiar?"

"I don't think Charlie wants to find us or us to find him. He knows we're leaving and has nothing to gain by making contact."

"My point exactly. The Indians were not interested in making contact with the Army either."

"Well, LT, it sounds like the same shit, different war."

SGT Michael "Stringbean" Stringfellow

Landing in Elephant Valley

18. DOMINOES

The low hill country in which we were now working contained numerous streams and trails coming out of the mountains. Our job of interdicting the flow of VC and NVA to Da Nang and its surrounding villages involved patrolling these thoroughfares, looking for signs of recent activity and, if we were lucky, catching the enemy off guard. While the vegetation on the foothills was sparse and scrubby, along the streams and rivers it was lush and thick.

With my map spread out on the ground, I gave my orders for the day's activities:

"See this blue line and how it forks? Rod, I want you to take your squad to the northern fork and work your way back to the west. We're looking for trails, base camps, caches, any sign of the dinks."

"Chris, I want you to work this southern branch from east to west and look for the same stuff. I'll be going with you, and I'll walk third in line."

"Stringbean, you're getting over today because you're going to stay here and secure the NDP."

"We'll be going light so take what you need and leave the rucks in the NDP. Report to me anything that you find. If either of you get close to the stream junction, hold it up and call in. We don't want to fire each other up. Any questions? OK. We'll move out at zero eight hundred."

We covered the five hundred or so meters to the stream in good time, in spite of the fact that it was already very hot. Granado, who was walking point, reported to Sergeant Christiansen:

"There's a dink trail next to the stream."

From a short distance, all that could be seen was a wall of bamboo paralleling the stream. What Granado had discovered was a hidden trail on the other side of that wall.

Sergeant Christiansen said, "Whaddaya think, LT?"

"Let's go and have a look."

Decisions like this were the leader's dilemma. In training, we were cautioned to never walk on a VC or NVA trail. At the same time, we were also told that sometimes there was no other choice. In this case, cutting a trail through bamboo was out of the question. Trying to cut bamboo is like trying to cut through steel. It makes a lot of noise and it quickly exhausts whoever is cutting trail. In short order, the entire squad would be worn out. Besides, our mission was to find signs of enemy activity.

"OK, let's follow it but be damned careful. Take it slow and make sure Granado keeps an eye out for trip wires."

We squeezed between the bamboo and onto the trail and in doing so entered another dimension. The Vietnamese are generally very short and this trail was carved out of the bamboo in order to accommodate them, not us. We found ourselves in what amounted to a bamboo tunnel that was about four and a half feet tall and wide enough for one man. The tamped down path made it obvious that this trail was frequently and recently used. It was very dimly lit, with the sunlight filtering through the thick foliage. The stream had disappeared on the other side of the tunnel wall. With no air circulation, it was like being fully clothed in a sauna. We crept along in complete silence, bent over at the waist and drenched in sweat. Wondering what lay up ahead, my breath came in short pants and my heart was pounding. Granado was out front on point, followed by Sergeant Christiansen as slack man and the connecting link between the point and the rest of the squad, and then me.

!!!BAM!!! One instant I was looking at the backside of Sergeant Christiansen and the next instant he came crashing into me, knocking me backwards into the man behind me, and we all started falling over like dominoes. As soon as I regained my composure I crawled forward to find Granado untangling himself from Sergeant Christiansen:

"What the fuck happened?"

"I came around a bend in the trail and there was a huge crocodile there."

"Are you shitting me?"

"No, sir."

"What was it doing?"

"I don't know, I got the hell out."

"Alright. Let's move out. Take it real slow, Granado."

We proceeded on until we eventually emerged at the base of the more heavily vegetated mountains. There was a large pond there and we could see other crocodiles in the water.

"And thrice he heard a breech-bolt snick tho never a man was seen."
Rudyard Kipling

Meanwhile, Sergeant Rodriguez called me on the radio.

In a stage whisper, he said, "We have been moving up the blue line. The point heard three clicks like weapons being switched off safe. We are in the water and exposed, so we backed off."

Sensing that this might be our big chance, I responded, "Roger, go ahead and fire it up. I'll come with this element ASAP."

Shortly we heard the sounds of automatic weapons fire interspersed by the crump of grenades.

It took us a while to backtrack and make our way to the other branch of the stream. There being no trails, we, like Rod's squad before us, waded in waist-deep water until we caught up with them. They hadn't received any return fire so they cautiously moved in on the location from which they had heard the noise. It was a cleverly concealed fighting position with an NVA canteen and a couple of ponchos left behind. I was very impressed with the workmanship of the position. It was reinforced with hand-hewn beams which were joined to the side beams with dovetail joints which were also hand hewn. These guys were no slackers.

If I hadn't been so gung ho and so eager to engage with the enemy, I might have indulged in some sober thoughts on how many times so far we might have had our asses handed to us.

19. FIRST CONTACT?

The thirteenth of February, the thirty-sixth day of this mission, found us again in the low, hilly area at the base of the mountains. I was with Sergeant Christiansen's squad patrolling along a large stream. We had received intelligence reports of extensive enemy activity here and the possibility of weapons caches. Hiding weapons along the bank of a stream or even sinking them in the water was fairly common practice by the VC.

After over a month of experiencing not much more than rain, heat, leeches and ticks, this had the makings of yet another fruitless walk in the hot sun. This low expectation was offset by the fact that we had received specific reports concerning this area, so we were attentively examining every inch of the stream bank and bed as we walked along the water's edge.

Out in front of me was Warner on point, Raley the scout-dog handler along with his dog Von, and then Sergeant Christiansen. Because the terrain was fairly open, we were able to keep a good fifteen meters between us as we moved in single file. Just as I was moving out of some chest-high reeds onto a sand bar, my attention was seized by an image that will forever be etched in my memory. Von was lunging toward the bank in a personnel alert posture and Raley was standing straddle-legged firing his carbine on full automatic at a clump of bushes on the other side of the stream about twenty meters away. The next thing I knew, I was in the prone position firing and throwing hand grenades as were Warner and Sergeant Christiansen. In the only out of body experience I have ever had, I was engaged in this frantic physical activity while simultaneously looking down at myself from above and marveling at how fast I had gotten down and started shooting.

After our initial burst of fire, Raley, Christiansen, and I rushed forward on the sand bar to get into a better firing position. Now it was time to exercise my role as a leader. I whirled around to give instructions to the soldiers who were behind me and…. No one was there! The soldiers immediately behind me were brand new replacements who had gone to ground and were hiding behind a sand drift now about forty meters to my rear. While I was gesturing them to come up, Bugge, my red-headed hippy grenadier from San Rafael, California, along with a couple of the other old hands, took the initiative and started across the stream in a flanking maneuver.

It soon became apparent that we were not receiving return fire so I called out, "Cease fire, cease fire."

Chris and I approached Raley, "What the hell happened?"

"I saw movement in the bushes just as Von alerted and I wasn't about to wait around to find out what it was."

"I saw movement too," Warner added.

"OK. Let's see what we've got…Bugge," I shouted across the stream, "Whaddaya got?"

Bugge and his crew were now cautiously approaching the bushes and Chris and I back-tracked to cross the stream at the same point where Bugge crossed.

"There ain't nothin' here."

The day immediately reverted to its former status as a hazy, lazy, hot sunny day.

Again, nothing. Within the clump of bushes there was a hide position and several enemy trails leading away to the south. We were foiled again in our quest to "close with the enemy by means of fire and maneuver in order to kill or capture him."

Still, it hadn't been all bad. One of the biggest things a soldier, especially a leader, wonders and worries about is how he will react to the first taste of combat. I didn't panic, I didn't freeze and for that I was satisfied with myself. No, I was proud of myself.

Back on the north side of the stream, Chris got the now quite jittery squad reorganized to move out. Everyone seemed apprehensive of what lay ahead. Just at the point where we were ready to move, Warner announced, "I don't want to walk point."

In infantry officer training they try to prepare you for every eventuality and how you should react to it, and this situation was certainly one that would have been covered. However, what you say you are going to do in a training situation isn't necessarily what you are going to do when it's the real thing. Should I order him to walk point? Should I threaten to court martial him or perhaps shoot him? Should I publicly shame him in front of the rest of the soldiers by calling on someone else? What do I do if no one else will do it? This was Warner, one of my best soldiers who had already logged two or three years in Vietnam as an infantryman.

One of the things that a leader (this leader anyway) learns in combat is that compromise is inevitable. Power and authority exist only to the degree that subordinates are willing to submit to it.

"Fuck it, I'll walk point," I said.

I walked point for an hour or two and then Warner caught up to me and said, "I got it, LT, I'll take it from here."

"You sure?"

"Roger that."

"OK. You got it."

Later that day we took re-supply. Between being engrossed with letters from home and the Stars and Stripes newspaper you'd have never known we'd had the drama of the morning. Human beings are adaptable creatures.

SP4 Railey and Von

20. RIDGELINE

We spent the next week humping and patrolling by day and putting out ambushes at night, with nothing to show for our efforts.

Our next assignment was to man The Ridgeline for ten days. The ridgeline was, as its name implied, a ridge located North of the city of Da Nang. It was anchored on the West by Hill 327, where there was an observation post, and from there it stretched to the East for approximately four kilometers. There were two-man bunkers every twenty meters or so. It took four companies to man it.

This would be a mixed blessing. On the one hand, we would have a break from the physical hardships and dangers of the field, and we would receive two hot meals per day. It was relatively safe although it had been attacked in the past. On the negative side, it was going to be boring, with little to do other than fill sandbags, string barbed wire, and go on short patrols forward of the wire.

Being in command of bored soldiers was a major concern for me. In the field, no one used drugs, mainly because they were viewed as jeopardizing safety and no one wanted to be the last soldier killed in Vietnam. But, when soldiers felt secure, they saw no harm in smoking marijuana and, of course, there were much stronger drugs readily available in the rear. Feeling secure and being secure are not necessarily the same thing and I knew I would have my work cut out for me.

On the 20th of February, we were extracted from the field. I, along with everyone else was looking forward to a shower and some clean clothes, which we had not felt for two weeks. Keeping to the dictum that leaders are the first in and the last out, I was on the last bird back to the base camp. When I arrived, the XO, Jim Cole, was at his usual place on the edge of the chopper pad rodding weapons as we filed off but, instead of filing back into the base camp, we were being herded onto deuce-and-a-half trucks.

"Jim, what the hell is going on?" I asked.

"Battalion wants us to move straight up to the Ridgeline."

"Are you shitting me? No stand-down, no showers, no change of clothes?"

"I'm afraid not, Buzz. It's not my idea."

"Well that's really fucked up. This is like that old Army game, 'You play ball with me and I'll shove the bat up your ass.'"

"Yeah, it feels like that, but there it is."

"Well fuck it then, it don't mean nothin'."

Most of the platoon was already on the trucks and the remainder were climbing aboard, so I jumped in the cab of the lead deuce and a half and we headed out. Once through the gate of Camp Crescenz, we turned right in the direction of the brigade headquarters. To the right of the road the land was flat—rice paddies stretching east to Da Nang and its large harbor. To our left the land rose steeply up to the ridge whence we were headed. After about a mile we turned left onto a dirt road, which led up to the top of the ridge. The trucks shifted into low gear as we ground our way to the top in a cloud of dust.

"How're you doing," I said to the truck driver.

"You know how it is, sir, same old shit, different day. How're you doin'?"

"I'm doing fine. Every day's a holiday and every meal's a banquet. What's the word in the rear?"

"The word is that we're standing down in April. That would be some good shit, huh, sir?"

"Well, you know, one of the first things I learned in the Army was, 'What you see is what you get.' I'm not holding my breath for anything. When we're told to stand down, I'll know we are standing down. Until then we'll just drive on."

"Yeah, I guess that's about right. But I'm sure looking forward to getting back to the world. I see those Freedom Birds leaving every day and I sure wish I was on one."

"Don't worry, you'll be on one sooner or later."

Once at the top, the trucks turned left onto a road which ran the length of the approximately four-kilometer ridge, culminating at its Western end with the dominating presence of Hill 327, home of the command group which had charge of the four companies manning the defenses. The top of the ridge was completely bare and the slopes were covered with nothing more than scrub brush, allowing for observation in all directions. Off the Northern crest were bunkers located every twenty or so meters for the entire length of the ridge. For the next ten days, a three-hundred meter segment of this would be our home.

As the trucks came to a halt, I dismounted and went back to the second truck, where Sergeant Rodriguez was riding in the cab.

"Get everyone off the trucks. Put them in squad groups off to the side of the road and have them hang tight until we find out what the fuck we're doing up here. Then catch up to me. I'm going to meet the platoon leader of the platoon we are relieving."

We were replacing one of our sister battalions, the 3d Battalion of the 21st Infantry Regiment, commonly known as the "Gimlets". I joined Captain Thomas as we met with our respective counterparts for a tour of our sector.

As I approached my counterpart, I immediately recognized him as Brian Campbell, my buddy in Ranger School.

"BC! How the fuck you doing?"

"Hey, Buzz. I didn't know you were over here."

"I didn't think any of us were coming over, but when all you WOOP's (West Pointers) left in August, I decided I couldn't let you have all the fun, so I called Infantry Branch and demanded my right to my choice of assignment as a Distinguished Military Graduate. And here I am."

Meanwhile, Sergeant Rodriguez had caught up with us.

"This is my Platoon Sergeant, Sergeant Rodriguez."

"Good to meet you. This is my platoon sergeant, Staff Sergeant Reynolds. Welcome to the Ridgeline. You'll be taking over fifteen bunkers, usually five bunkers per squad. As much as possible I assign two people per bunker. C'mon and I'll show you what you've got."

We started at the western-most bunker, where I would be tying in with Second Platoon, and worked our way east, inspecting each bunker along the way. I made a mental note that the bunkers had been cleaned out and were being handed over to us in good condition. Each position had a sleeping bunker dug into the face of the ridge and covered with a galvanized steel culvert half which was then covered with several layers of sand bags. In front of each sleeping bunker and slightly further down the slope was a fighting position, again with overhead cover and heavily sand bagged parapets. In front of the fighting positions were several coils of concertina wire.

"Have you had any contact up here?"

"No, not really. No firefights at any rate. You've got to be careful though because there are beaucoup trails leading up here from down in the valley and we did detect recent activity on them. It's pretty quiet but you can't afford to be half-stepping."

About mid-way along the line we came to Tower 14, which would be where I established my CP. We climbed up into the tower and Lt Campbell oriented us to the surrounding terrain.

"Due West, right there, is Fire Base Maude and behind it the Maude AO."

"Yeah, that's where we just came from."

"And down there, to the Southwest, is Fire Base Linda and the Linda AO. That ridge there on the left is Charlie Ridge."

Way off in the distance, looming ominously under heavy clouds was the dreaded Charlie Ridge, home turf of the NVA and the site of many battles. We knew we were scheduled to head back there when we finished our stint on the Ridgeline.

"Rod, you guys were on Charlie Ridge recently, weren't you?"

"Yes, sir. We were up there in October."

"Isn't that where you got into a big fight with the NVA and the platoon sergeant got killed?"

"Roger that. Sergeant Thompson and Granado were up on point when all of a sudden, Sergeant Thompson yelled, 'Contact!' Then he got shot right between the eyes. Granado was stranded up there with him until it was over."

"No wonder no one wants to go back there."

We continued our tour of my new area of responsibility until we reached my right flank, where I would tie in with Delta Company's left flank. I thanked LT Campbell as he headed off to organize his platoon to vacate as we moved in.

"Good luck, BC. It was great to see you. Keep your head down."

"Thanks, Buzz. See you around. Good luck to you, too."

Turning to Sergeant Rodriguez, I said, "OK, Rod. Let's keep it simple. First Squad will take bunkers 1-5, Second Squad bunkers 6-10, and Third Squad bunkers 11-15. You and I will occupy Tower 14 and the bunker that goes with it. Tell the squad leaders to take care to put their grenadiers where they can cover dead space and the pigs (machine guns) where they have the best fields of fire. We also have to make sure the pigs have interlocking fields of fire. Before everyone gets completely settled in, you and I will check the emplacement of the weapons. Any questions?"

"No, sir."

"OK. Get the squad leaders putting their squads in. Once they are there, you start walking the line from the West and I'll start from the East."

"Roger that, LT."

1st Platoon sector of the Ridgeline

LT John Beasley, D Company

21. SETTLING IN

The first order of business was to make sure we were ready to defend our position and to convey to the platoon that, even though we weren't out in the bush, this was still serious business. Unlike our normal active patrolling operations, this was a classic, by the book, static defense. Before commencing my inspection of the platoon sector, I reviewed in my mind the key elements that I should look for.

Each squad had an M-60 machine gun, or "pig", which was its main source of firepower. That weapon should be positioned where it could deliver flat trajectory fire on the most likely approach into the squad sector. It should also be positioned where it could cover as much of the squad frontage as possible. Next, the M-203 Grenade Launcher, or "Blooper", provided the squad with indirect fire and should be positioned to cover "dead space" like depressions or draws that could not be covered with flat trajectory fire. The fires of positions on the left and right flank should interlock with the fires of the squads to the left and right so that there would be no gaps. The squad's fire plan should be drawn out on something weatherproof, showing each weapon and its fan of fire. Dead space should be identified and shaded in.

Sergeant Christiansen's squad was the first that I checked.

"Sir, here is my squad fire-plan. I've got Specialist Tetrault in the center with the pig where he has good fields of fire across the whole squad sector, and he can also interlock fires with the squads on my left and right. There's a draw leading downhill from this position and I've marked it as dead-space on the fire plan as you can see. I've got Brown covering it with the blooper. There's also some dead-space here which we can cover with hand grenades."

"It looks good, Chris. Did you walk the dead-space? "

"Yes, sir. Both Brown and I walked it. We put some trip flares down there and, when we get some wire up here, we'll put in some tangle-foot and concertina."

"Outstanding. Hey, what do you think about putting Brown up for sergeant next time there's a promotion board? I've got Shake n' Bakes coming out my ass but most of them aren't worth a shit. I'd much rather promote guys that have been here for a while and know their shit."

"I agree, LT. I think Brown would make an outstanding NCO. So would Elzey."

"Yeah, I've had my eye on him too. Listen, be thinking about a roster to get your guys back to the rear one at a time to get a shower and a change of clothes. After I finish checking the positions, I'll have Rod set up a rotation plan with the squad leaders."

"Roger that, LT."

"OK. Your sector looks good. Good work."

I moved on to the next squad sector where I met Sergeant Blankenship (pseudonym).

"Good morning, Sergeant Blankenship, are you ready to show me what you've got?"

"Yes, sir."

"Let me see your fire plan."

The first clue that this wasn't going to be a repeat of Sergeant Christiansen's performance came when Sergeant B handed me his supposed fire plan. Instead of being on a large, durable piece of cardboard from a C-ration case, it was scrawled on a tiny piece of note-paper from a pocket notebook.

"What's going to happen to this the first time it rains?"

"Uh, I'll keep it in my pocket."

"That's the wrong answer. You need to have it where it can be found and used by anyone who needs to know what's going on here. This is useless. Where do you have your 203 (grenade launcher)?"

"Over there at the next bunker."

"I can see from here that there's no dead space in front of that position and I can see a big draw right in front of the position where we're standing. Did you get out in front of your position and walk the terrain?"

"Uh…"

"God-damn it, Sergeant B, you need to get this sector squared away. Put your pig where it has the best field of fire, put your 203 where it can cover the dead space, and fill in the rest with your riflemen. Walk the terrain to your front and mark dead space and obstacles. Shoot azimuths for each weapon and mark them on your fire plan. And get a piece of C-ration cardboard that won't melt the first time it rains. Do you understand?"

"Yes, sir."

"I'll be back to check and you better have your shit together."

"Yes, sir."

The next squad, while not as good as Sergeant Christiansen's, was satisfactory.

Back at the platoon CP, I gave Sergeant Rodriguez further instructions.

"We've got permission to send three men per day to the rear to get a shower and a change of clothes. We'll send one man per squad every day. Have each squad put together a roster so that everyone gets a chance to go back. Stay on Sergeant B. Man, he is a sorry excuse for an NCO. Let me know when he is ready for me to re-check his fire plan."

And so we settled in to a routine of watching, waiting, filling sand bags, patrolling, contending with wildly fluctuating weather and, above all, boredom.

22. NEW BATTALION COMMANDER

The first few nights on the ridgeline were pretty spooky on account of the mongooses and the 'fuck you' lizards. The mongooses rummaged around in the concertina wire at night, rattling the noisemakers that were tied there, and the 'fuck you' lizards made an eerily human call that sounded just like their name. I inspected the platoon sector each day and went on short patrols in front of our positions to check for enemy activity. The cat and mouse game with Charlie continued. It wasn't unusual to find places where the wire had been cut or pushed down, and trip flares disconnected. To counter that activity, I directed that Mechanical Ambushes be emplaced where tampering was detected.

The weather on the ridge was windy, often foggy and cold. The swirling fog added to the eeriness. In addition to our sleeping bunker and fighting position, Sergeant Rodriguez and I also occupied a forty-foot high observation tower, which shook and howled in the wind.

The whole company sector was wired into a field telephone network so that we wouldn't have to rely exclusively on radios and their requirement for brevity and coding and decoding messages. A couple of days after our arrival on the ridge, the platoon leaders received a net call from Captain Thomas. After each of us had answered the call, he announced: "Looks like we have a new battalion commander. His name is Colonel Andrew J. Perkins. Over the next few days, each of you is going to go to the rear and have dinner with him so he can meet all his officers. Lieutenant Sherwood, you're going to be first, tomorrow night."

"Aw, why me?"

"Because you're the 1st Platoon Leader and you're going to be first. You have a problem with that?"

"No, sir."

"I didn't think so. Be ready to go to the rear at 1700 hours. Salazar will run you down in my jeep." Salazar was one of my soldiers who doubled as the CO's driver when we were in the rear.

"Roger that, sir."

At the appointed time, Salazar delivered me to the Camp Cresencz mess hall, where I linked up with the other lucky lieutenant who was to dine with the new colonel. This was to be the fourth battalion commander under whom I had served in my less than two years in the Army. Colonel's Simmons, aka 'Smoke', Gilmore, and Tate all had different personalities, but all were cut from the same bolt of cloth: they were hard, tough infantrymen with extensive combat experience in Vietnam.

"Good evening, sir. Lieutenant Sherwood, Charlie Company."

Those were the words out of my mouth, but in my mind I was thinking, 'You gotta be kidding me!'

Before me stood a slightly pear shaped man who would best be described as portly. On his upper lip was a pencil thin moustache, which hearkened back to an earlier era, never mind that the wearing of moustaches by officers was highly frowned on. He looked out of place in jungle fatigues. I could better picture him in a suit, selling used cars or running for some small-time political office. I noticed that he was not a Ranger and, AND, HE DID NOT HAVE THE CIB! (The Combat Infantryman's Badge, awarded to all infantry soldiers who have served in combat). How could this be? How could an infantry officer not have served in combat after six years of this war? This did not inspire confidence.

For all of that, he was a nice enough human being, and dinner passed pleasantly if not memorably. Following dinner, Colonel Perkins invited us to join him at the evening Command and Staff Briefing, referred to derisively as the 'Lifer Meeting'. How could we say no?

The Command and Staff Briefing is one of the ways in which a commander is routinely kept up to date by the various members of his staff (S-1—Personnel and Administration; S-2—Combat Intelligence; S-3—Operations and Training; S-4—Supply and Logistics and S-5—Civil Affairs). Holden Caulfield had nothing on me with his disdain for anyone older than him. I observed the whole affair with mocking humor, sneering inwardly at each staff officer who got up and gave his ass-kissing briefing to the new commander. The whole affair was orchestrated by Major "Bruiser" Lenhart, the Executive Officer, whose own ingratiating comments were periodically inserted.

Nothing lasts forever, and the briefing mercifully came to an end. Salazar delivered me back up to the ridgeline around 2100 hrs.

23. BOREDOM AND DRUGS

The next day I got another unpleasant surprise from Captain Thomas.

"You're getting a replacement later today. He just got released from the drug detox center in Cam Ranh Bay. He's being processed out of the Army but he's going to be with you until that happens."

"You gotta be shittin' me, sir. They ought to lock these fuckers in a CONEX container and throw away the key. Or at least hold them in the rear. Why the fuck are they sending them out to us. That's bull-shit."

"Maybe so, but you're getting him."

My new charge showed up on the chow truck that brought us the noon meal. His name was Thurman. Unlike the crazy characters I had seen coming out of the detox center when I first arrived in country, Thurman was subdued, even polite.

"So what have you been taking? I know you weren't sent to detox for smoking grass."

"Heroin."

"Damn! How did you let yourself get hooked on that shit?"

"I don't know, sir. I've been on it since I was seventeen—before I came into the Army."

"Are you serious about staying off it?"

"Yes, sir. Definitely."

"Well, OK. Look, I'll help you in any way I can and I want you to hang on until you're sent home. I want you to succeed and I'm willing to help you, but I also want you to know that if you try any of that shit up here, I'll burn you. Understand?"

"Yes, sir."

"I'm going to assign you to Sergeant Christiansen's squad. He's my best squad leader and he'll treat you right if you treat him right. OK? I'll have Sergeant Rodriguez run you down to him. You should be able to catch them in time for chow."

Thurman wasn't the only drug problem that I had. As I had feared, being in a static position created a false sense of security which, less than a year previous, had resulted in this same company getting overrun on Fire Base Mary Ann, with 33 KIA and 82 WIA.

One morning as I was walking the bunker line, I discovered a large, one-quart plastic bag of marijuana sitting in plain sight in front of Specialist Bourland's sleeping position. Bourland had recently joined the platoon as an in-country transfer from up around Phu Bai. He looked squared away and outwardly played the role, but there was something about him that seemed kind of shifty. I found him nearby and confronted him.

"Bourland! What the fuck is this?"

"I don't know, sir."

"What do you mean, you don't know? It was sitting right in front of your sleeping position."

"I don't know, sir."

In order to charge him with drug possession, the drugs had to be under his control. Since they were sitting out in the open, they were, legally speaking, not in his possession.

I didn't know quite what to make of it. He had to know that I would see it. Was he testing me to see how I would react to grass being up on the ridgeline? I would never know, but one thing I did know was that I wasn't having it.

"Well, here's what let's do. Come with me."

Carrying the bag of marijuana, I had him accompany me down in front of his fighting position.

"You see this big bag of grass that you say you don't know anything about? Watch this."

With that, I scattered the contents in the concertina wire.

"Don't let me catch you with this shit or I'll burn your fucking ass."

At my next meeting with Sergeant Rodriguez and the squad leaders, I reiterated my feeling about drug use, knowing that they would be my most effective tools for preventing it. I never had any further issues with drugs on the ridgeline, but I was never as confident as I was in the bush that they weren't there.

Laziness, including my own, was also an issue. There just wasn't that much to do and, within a couple of days of being there, I was bored stiff and felt lazy into the bargain. Some days I felt so lazy that I didn't even want to climb down out of the tower to take a leak.

One day when our main task was to fill sand bags, I was walking the line checking on work. I came on Sergeants Mixon and Willard sitting idle next to a pile of unfilled sandbags. (Sergeant Mixon was the Airborne Ranger that I was so pleased to have gotten).

"What are you guys doing?"

"Taking a break, LT."

"Bull-shit! You're not doing a fucking thing! I'm not stupid. There's not a sand bag in sight that's been filled."

"Aw, come on sir, it's too hard."

"My ass. You guys are NCO's. You're supposed to be setting the example. Come on, let's go. Let's fill sand bags. Hold it open."

I grabbed the entrenching tool and started shoveling dirt into the sand bag. I shoveled for about fifteen minutes.

"See, it's easy. If I can do it, you can do it. Now you shovel."

For several more minutes, I held the bags open while they shoveled until I felt they had the rhythm and the message.

"All right. You got it? That's what I want to see. Hubba, hubba."

A piece of good news came to me when I got word that my buddy, John Beasley, was being assigned to D Company of our battalion. I got along well with the other officers but it would be nice to have a good friend in the unit. I very much looked forward to re-connecting with him. Meanwhile, my main task was to fight my own boredom by keeping myself and the platoon as busy as possible.

24. MORE RIDGELINE AND A DINING-IN

Officially, Vietnam had entered the hot, dry season, but you would never have known that on the Ridgeline. We'd been up there for a week and every day it got colder and colder and a hard wind constantly blew. I even retrieved my field jacket from my duffle bag in the rear and took to wearing it as I made my rounds. The only thing that could make it worse, I thought, would be if it started raining.

The bright side of things was that John Beasley had returned from R & R and had been assigned to D Company of the battalion. Not only that, his platoon was on the left flank of D Company and my platoon was on the right flank of C Company, which meant that our platoons were side by side and we would be able to visit frequently.

Another bright spot was a large "cush package" from my wife. It was full of chocolate chip cookies, broken up and stale but delicious nonetheless, cans of pudding and fruit and a large bag of assorted dried fruit. Besides the home baked cookies, my favorite was the dried fruit. On returning from my rounds one day, I climbed up into my tower, relishing the thought of the tart chewiness of a dried apricot, my particular favorite. I could already taste it. I greeted Sergeant Rodriguez and went immediately to the bag of dried fruit on my cot and reached inside. There were dried peaches, dried pears, and plums, but no apricots.

"Hey, where's all the apricots!?"

"I ate them."

"You ate them? You ate all the apricots? Those are my favorites. You fucker!"

"I couldn't help it, LT, they were so good." As he said this he burst out laughing.

"You fucker!"

Even though I was angry, I couldn't help but see the humor in the situation and I was both laughing and pissed off all at once.

"I'm going to eat all your fucking jalapenos next time you get a care package!"

That made Rod laugh even harder.

The next day I wandered over to D Company to visit with John Beasley. "So how does it feel to be back in a unit?"

"It's pretty good. Being in Hoi An was pretty boring…and I was lucky to even have that assignment. A lot of the lieutenants from the 101st ended up in Saigon warehouses counting socks, air-conditioners, inspecting weapons, and stuff like that."

"Ooooh! This is boring enough. Having a job like that would be awful. We're supposed to be up here for eight more days and I don't even know if I can stand that. I'm really looking forward to getting back to the bush, but at the same time I'm getting a little nervous about it. We're going to Charlie Ridge and that's where all the contact has been. I've been trying to get my guys keyed up for it. I wonder when we're going to stand down?"

"That's the big question. I wouldn't be surprised if we didn't make it through the next mission. I'm sure they'll keep us in the field right up until the last minute while the rear is taking care of all the paperwork."

"Well, I'd better get back to my platoon. You going to the dining-in?"

"Yeah. It should be fun."

"Good. If I don't see you before, I'll see you there."

A couple of nights later, all but a skeleton crew of officers were taken to the rear for the dining in. A dining in is a formal dinner for men only (a Dining-out includes the ladies), presided over by the commander and packed full of formality and tradition. Back in the States the dress blue uniform would be worn, but here it meant your best jungle fatigues and boots.

There were many toasts: to the Commander in Chief, the U.S. Army, the 46th Infantry Regiment, the ladies, etc. I was pretty well lubricated by the time the main course was served—steak and lobster. Sitting next to me was Lieutenant Lynn Rolf, a Mid-Westerner from Minnesota. Out of the corner of my eye, I noticed that he seemed a little hesitant about his lobster. Sensing an opportunity, I attacked mine with gusto. Having devoured the main body, I picked up the head and began sucking the fats and juices out of the head cavity with great slurping noises, just as I would a crawfish back in Louisiana. Peeking again at Lynn, I noticed that he had still not touched his lobster. In fact, he was looking a little green around the gills.

"Aren't you going to eat your lobster?" I asked innocently.

"Uhh, I don't think so. Do you want it?"

"Hey man, I don't want to take your lobster."

"No, really, go ahead."

"Well, if you insist, I won't say no."

The red monster was transferred to my plate and I wasted no time getting it tucked in, lest my friend change his mind.

After the formalities of the dinner, we were treated to a floor-show. More Koreans singing, "To dweam the impossible dweam." It wasn't too good, but any entertainment was better than no entertainment.

A couple of days later, I received orders for the coveted silver and blue Combat Infantryman's Badge (CIB), which is awarded to infantrymen for combat service. This was indeed something I had coveted for years, but now, receiving it was anticlimactic. Maybe that was because of the boredom and blahs with which the Ridgeline had infected me. Six more days to go.

25. I CONTEMPLATE MURDER

I had been systematically sending my men to the rear on a one-day pass in order to shower, relax, and maybe even take a jaunt to the Air Base. Satisfied that I had taken care of my men, I decided to take a pass myself. I proposed to my buddy Beas that he do the same and he agreed. So, on 8 March, we rode the morning chow truck to the rear, changed into our best fatigues and caught a ride to Gunfighter Village on the Da Nang Air Base.

As we entered the main gate and returned the salute of the Air Policeman (AP) on duty, we were immediately transported into another world, a world we had seen before, but seemingly a lifetime ago.

"Wow! Look at this, will you?"

"Man, these Air Force guys have got it made. Shit, they've even got air-conditioning."

"Well, you know the Air Force isn't even a military service. They're just a bunch of civilians who wear blue suits."

Arrayed along the street along which we walked were three-story cinder block barracks buildings featuring individual rooms, each with its own air-conditioner. We were like country come to town, gawking at the big city sights. To our left was the Post Exchange (PX) complex, including the PX itself, (an all-purpose department store typical of any large military base in the States), assorted snack vendors, clubs for the enlisted men, the works.

"Hey, I heard that they even have a restaurant here."

"No shit! Let's find it."

It didn't take long before we found the Bamboo Room, which served Asian food. Still feeling like country bumpkins in the big city, we were amazed to be greeted by a hostess who seated us at a table and presented us with a menu with a variety of dishes to pick from.

"Ah you weddy to ohda?" asked the Vietnamese waitress.

"I think I'll have the shrimp fried rice."

"The same for me, and bring us a bottle of champagne."

Over shrimp fried rice, (which was about as far as either of us was willing to venture in the Asian food department) and champagne, John and I reminisced over all the funny people and events from our previous assignments together. Colonel "Smoke", Sergeant Major Sabalauski, Major "Peanut Balls" Carey, Captain Ferguson, Warrant Officer Nikolai, and, of course, the time John fought one of the soldiers behind the Motor Pool.

"You remember the time Leal was dogging it on push-ups during PT and you told him if you were a private you'd kick his ass?"

"Yeah, and he challenged me."

"So what ever came of it?"

"We met behind the Motor Pool and, as soon as we squared off, he rushed at me and he slipped and fell and broke his arm, so we never did actually fight."

"Did Smoke hear about it?"

"Yes, but he was such a wild man, he thought it was pretty cool. If it had been anyone else, I could have been in big trouble."

"You know, Leal worked for me when I was the Motor Officer. I'd watched too many Sergeant Bilko shows when I was a kid, so whenever we needed a part, I'd just have him go and steal it off another vehicle."

"Those were some wild times."

I got the attention of the waitress.

"Can you bring us another bottle of champagne?"

The more we ate and drank, the further down memory lane we wandered.

"Remember when we got to the Florida phase of Ranger School and Colonel Tucker said he would accept two or three dead Rangers per class in order to make it more realistic?"

"Yeah, he was a crazy son of a bitch. He almost got his wish, too, that night we crossed the swamp and the Yellow River. I have never been so cold in my life. Whenever I lifted my arm out of the water, the water on my jacket immediately turned to ice."

"We were in it all night. You didn't know you'd hit the river in the middle of the swamp until the point man went under water.

"My patrol made it to dry ground right at dawn. As soon as the guy in front of me put his feet on dry ground, he fainted. He went down like a tree, flat on his face."

"That was a pretty tough night. You about ready to head on back?" John said.

"I think I'll hang around here for a little longer. I'm going to poke around the PX."

As we left the Bamboo Room, John turned back toward the main gate and I headed back to the PX complex.

"OK, I'll see you in a bit."

After looking around the PX and finding nothing that interested me, I decided that I was still hungry so I headed back to the Bamboo Room. As I entered, I noticed that my heroin addict, Thurman, was sitting alone at a table so I asked him if I could join him. I ordered another meal of shrimp fried rice and, as I ate, I talked to him about his drug problem. He told me he was trying hard to get off it. He sounded truthful and had nothing to gain by lying to me since he was being discharged from the Army as unfit.

"So how do you get the shit?"

"The whores bring it. They sneak up at night to do their thing, but they also sell smack."

"Who do they work for? Who's the pimp and main pusher?"

"He's a village chief. He never comes himself, it's always the whores who bring it."

"I'd like to get that fucker. Here we are fighting his war for him and he stabs us in the back."

I was a zealot when it came to drugs, but I was also like an indulgent parent who is blind to the faults of their own children. If G.I.'s were doing something wrong, it had to be entirely the fault of someone else.

"Do you think he would come if there was a really big order for smack?"

"I dunno, probably."

"How big an order do you think it would take to get him to come?"

"Maybe a couple of hundred bucks."

"I want to nail him. Would you be willing to place a big order to get him to come up?"

"Yeah. Yeah I'd do that."

"OK, let's do this...."

The plan was that Thurman would put in an order for three hundred dollars worth of heroin that same evening, to be delivered the following day somewhere on the Ridgeline. When he arrived with the goods, we'd have a little ambush waiting for him. My plan was to capture him, but to shoot him if he tried to run. I was actually hoping he would try to run.

Thurman excused himself to initiate the plan and head back to the Ridgeline. I was still feeling hungry so I ordered another plate of shrimp fried rice, my third of the day. Eventually, I made my way back to Camp Crecenz and from there back up to the Ridge, where I briefed Sergeant Rodriguez on our sting operation for the following day.

26. WAR ON DRUGS

The following morning, Thurman showed me the delivery point, a small clearing off the backside of the ridge. Nearby was a huge boulder, which would be the perfect spot to set up our ambush. I got the .45 caliber pistol that we kept in the platoon, as well as a rifleman named O'Brien.

"OK look "OB", we're going to try and capture this fucker, but if he tries to run, we're going to nail him. Don't you do anything unless I tell you to. Got it?"

"Yes, sir."

Thurman was standing in the clearing and O'Brien and I were well concealed behind the boulder. I was panting, fast and shallow breaths in my fearful anticipation of what would happen next. The appointed time came and went and I began to wonder if we had been stood up, or if Thurman had double-crossed me. Then I heard the buzzing of a motor scooter and I tensed up for the moment of truth. Into the clearing came a motor scooter with, not the village chief, but two girls on board, dressed in loose fitting white blouses and loose fitting black pajama trousers.

"Shit!" I muttered to myself. It seemed the village chief was too canny to come himself. Nevertheless, we had to follow through with the plan. One of the girls produced several small bags of a white powdered substance, which was my cue to move.

"Let's go!" I said to O'Brien.

"Dung lai, Dung lai!" I shouted as I burst from behind the boulder pointing the .45 at the girls.

Terrified, they froze in place.

"Lai dai." I said as I motioned them toward me with the Vietnamese style palm-down wave of the hand while still pointing the .45 at them. O'Brien was next to me pointing his M-16 at them. Thurman faded from the scene while I took the heroin from them, being careful to hold it in their sight, thereby maintaining the chain of custody, as if such legal niceties counted for anything over there. I escorted them, still frightened out of their wits, up to the top of the ridge and onto the road. I got hold of Sergeant Rodriguez and instructed him to contact the MP's to come up and take custody of my two prisoners. While waiting for the MP's to arrive, some of the soldiers in the vicinity came over to see what was going on.

Sergeant Mixon said to me, "How come you didn't waste them, LT?"

"Why the hell would I do that? They're unarmed and they didn't try to get away."

"I wish I'd been there. I would have blown them away."

I immediately sensed that, in addition to being lazy, Sergeant Mixon was also a cruel man who would like to kill someone just for the sake of doing it.

"Don't even think about it, Sergeant Mixon. I'll be keeping my eye on your ass."

He gave me one of his surly looks, which I had gotten used to, and turned away. As an Airborne Ranger, he was proof positive that having the right credentials is no guarantee of superior performance.

By and by, the MP's arrived and hauled the girls away. I never heard any more about it. The girls were probably released that day and who knows what happened to the heroin. So ended, somewhat anticlimactically, my one-man attempt to eradicate the illicit drug trade in Vietnam. In retrospect, I'm grateful that the village chief didn't come. He saved his own skin and spared me from doing something that I would have regretted for the rest of my life.

27. FAREWELL TO THE RIDGE

I was standing in the tower looking blankly at the same scenery we'd been looking at for two weeks. The boredom was monumental and, to make matters worse, our mission on the Ridgeline was extended for three days. Sergeants Rodriguez and Christiansen were standing at the base of the tower.

"LT! There's trmffr obgthh yogmf hmffd!"

"WHAT!? SAY AGAIN!"

"THERE'S TRACERS GOING OVER YOUR HEAD!"

"SHIT!"

I grabbed my rifle and scrambled down the ladder.

"Are you kiddin' me?"

"No, there were green tracers going over the top of the tower."

"Could you see where they were coming from."

"No, it was too far away. Didn't even hear anything."

"Well ain't that a bitch. I get shot at and can't even shoot back."

Laughing, Sergeant Rodriguez said, "Maybe that village chief that you tried to bust took out a contract on you."

At that we all three started laughing.

"Well, that's all the more reason to un-ass this AO.

"You know we're going to Charlie Ridge."

"I don't care where we go, as long as it's away from here."

The next day, "un-assing" the Ridgeline began to look like a reality. In the morning, the other platoon leaders and I took a VR (Visual-Reconnaissance) of Charlie Ridge in a LOH (Light Observation Helicopter, pronounced: 'loach'). After spending most of my time foot-slogging, it felt very upscale to be zipping through the air in the flying version of a sports car. I came away with the knowledge that we were going into very rugged terrain, much more rugged than we had worked in previously. Beyond that, I didn't know much since I didn't know exactly where we were going or what our mission would be.

After the VR, the rest of the day was spent drawing supplies and equipment. The day was rounded off by a midnight guard shift. It was very windy and cold. Bundled up in my field jacket and using the dim, red-filtered light from my flashlight, I took the opportunity to read letters from home.

Back in the States, life went on for my family. My pregnant wife was growing ever larger with our first child as evidenced by the photographs and perfumed reports that were contained in her letters. My parents also were going through big changes in their lives. My mother was changing jobs, which was stressful and my dad was going through the painful process of liquidating his business. To all I wrote words of encouragement and endearment, but, in truth, my main focus was on the job at hand.

The tangible evidence that we were leaving the Ridgeline breathed new life into me. Even the three-day extension was tolerable, now that I could devote my energy to preparing for our next mission, the dreaded Charlie Ridge. As I made the rounds of my platoon positions, I tried to get a sense of their attitude about the next mission. As usual, these were as varied as the people in the platoon. Some were full of martial bravado.

"Hey, LT, we gonna go out there and kick Charlie's ass!"

"Con huevos, cabron! Anything to leave this pinche ridgeline."

Others, especially some of the more veteran members, were not so enthusiastic.

"Charlie Ridge is no picnic, LT. It's crawling with dinks and those yamas will kick your ass. Maude was a picnic compared to that."

Still others were ominously resistant to the idea of returning to the scene of their last big fight.

"I ain't goin' back there. They can put me in LBJ (Long Binh Jail) before I'll go back to Charlie Ridge."

"There it is. At least in LBJ you ain't goin' to get killed."

I had always heard from the old-time career sergeants that a complaining soldier was a happy soldier. Much as I wanted to believe that was true, these mutinous statements unsettled me a bit. What if they were serious? What if, tomorrow, they refused to go. What if it became contagious? What if… All I could do was remain forceful and positive and wait and see what happened.

During the day I received a replacement soldier and a new platoon sergeant. The soldier, another in-country transfer, was quickly assigned to a squad. I would soon regret that he joined us. Staff Sergeant David Gauthier was my new platoon sergeant. He was a seasoned veteran in his third year in Vietnam. Like Tetrault, the machine gunner, he was from New Hampshire of French Canadian descent. He cut a good figure with his rakish handlebar moustache. We quickly got acquainted and I got the feeling that we would be able to work well together. Sergeant Rodriguez, who had done such a great job as platoon sergeant, went back to leading his squad.

We turned over the Ridgeline to the unit that was replacing us and loaded trucks which took us back to our base camp, where we got rid of stuff we had collected on the Ridgeline which wouldn't be needed in the field. My trusty field jacket was among those items left behind — seldom needed and much too heavy and bulky to lug around in the field.

Rucksacks packed, weapons cleaned and oiled, tomorrow we were bound for Charlie Ridge.

28. CHARLIE RIDGE

Due to heavy overcast, the air-assault onto Charlie Ridge was scheduled for early afternoon, giving the fog a chance to lift. Apprehension and the gloomy weather mitigated against the usual light-hearted banter, but, much to my relief, the whole platoon was present and there was no talk of refusal. I was feeling tense and excited.

2d Platoon and the Company Command Group were to go in first, followed by my platoon, with 3d Platoon bringing up the rear. We were going into a one-ship LZ so the insertion was going to take a while.

Approaching the LZ on the first bird of my platoon lift I saw a soldier standing amidst blasted trees in a postage stamp of cleared jungle, arms raised, guiding the ship in. The jungle was especially thick and the hills steep, the most rugged looking terrain I had yet seen. The door gunners on either side of the helicopter visually scanned the tree line, fingers ready on the trigger of their pedestal–mounted M-60 machine guns. The deafening "WOP-WOP-WOP" of the rotor blade drowned out all other sound.

'Where are we going to land?' I wondered, as the ship eased into the narrow clearing, which was littered with tree stumps and felled trees.

My question was answered as the pilot came to a hover and the crew-chief flicked his thumb at us, indicating that we were to jump.

Stepping out onto the skid, I jumped, aiming for a spot free of stumps. I landed hard but my rucksack continued the descent, shoulder straps digging in and the weight dragging me to my knees.

Staff sergeant Gauthier and the next squad soon arrived and we waited at the edge of the jungle for the rest of the platoon.

The jungle teems with life. Monkeys inhabit the trees as do a myriad of birds. On the ground are a wide range of animals, reptiles, and insects. Due to its natural aversion to human contact, we seldom encountered any wildlife but knew it was there. Our main companions were those creatures for whom we represented a free meal, namely the ubiquitous mosquitoes, leeches, and ticks. Like the city, with its constant thrum of ambient noise, there was always the sense of being in the midst of something alive. But not today.

Once the sounds of the departing aircraft faded to nothingness, an eerie, foreboding silence settled over the jungle, as if it were holding its breath in anticipation of something bad about to happen. Charlie was here and we knew it.

After waiting for several minutes for the rest of the birds to arrive, Captain Thomas called me over to where he was standing nearby. He said to me, "The fog has gotten too thick and the pilots have called off the insertion. We'll have to go with what we've got. I'm going to stay with 2d Platoon but I'm going to send Minh, the Kit Carson Scout[6] with you. We'll spend the night here and move out first thing in the morning. Maybe by then they'll be able to bring the rest of the company. Any questions?"

"Where do you want my platoon to go?"

"I want you to go south on that finger over there past the southeast corner of the LZ. Move up the finger to the top of the ridge and search

---

6. Kit Carson Scouts were former NVA or Viet Cong soldiers who changed sides and accompanied line units as scouts.

the area for dinks or recent activity. I'll be heading west from here with Second Platoon. That could all change tomorrow if the rest of the company arrives, but that's the plan for now."

"Roger that, sir."

"I'll send Minh along with you now. We'll form a tight perimeter around the LZ. You take the eastern half, from twelve to six. Second platoon will be on this side."

"Yes, sir."

Minh reluctantly left the CP group and joined me. He was visibly afraid, eyes wide and hesitant in all his movements. His behavior heightened my sense that the enemy was indeed nearby.

"You think VC here?"

"Yes, I think beaucoup BC."

"You walk point."

"I no want walk point."

Clearly Minh was going to be more of a liability than an asset so I decided not to push the issue of him walking with the point man.

I only had fourteen men on the ground, but it was plenty to fill our portion of the perimeter around the tiny LZ. Minh joined Staff Sergeant Gauthier, Gadzinski and me in the CP, which I established at the middle of our section of the perimeter. We maintained a high level of alert but the night passed uneventfully.

After the pre-dawn stand-to, we saddled up and moved out. Smitty was on point, about twenty meters in front, followed by Private Jeffries, then me and Gadzinski. As directed by Captain Thomas, we were moving south along a narrow finger that led up to the top of Charlie Ridge. It was heavily wooded with triple canopy growth and quite steep. The range of vision was no further than the man in front and the man behind. Nevertheless, we were moving as quietly as human beasts of burden could. We had been climbing for about fifteen minutes. Jeffries, in front of me, put up his hand signaling 'halt'. He turned his attention back to Smitty while the 'halt' signal was passed down the file. My attention was glued to the front.

"Shhhh." Jeffries signaled by putting his fingers to his lips.

"Dinks", he mouthed, pointing up ahead.

BANG, BANG, BANG. Three shots shattered the stillness.

The need for stealth and silence was now past.

"Let's go!" I yelled, as Gadzinski and I surged forward to where Jeffries was standing. I was excited. Finally, we were going to come to grips with the enemy.

"What happened?" I asked Jeffries.

"It was a trail watcher. He ran as soon as he saw us. Look, you can see where he was hiding."

I looked at the spot indicated. The foliage was tamped down and there were the remnants of a cigarette that he had dropped when he left the scene.

"SMITTY! What happened?" I yelled to Smitty, who was out of sight but within earshot.

"He ran. I got off three shots and then my weapon jammed."

"Which way did he go?"

"Up the hill and off to the left."

Gadzinski and I continued to move up to where Smitty was waiting. Inside myself I was very disappointed that Smitty's marksmanship hadn't been better, but I wasn't there and couldn't judge. I was still pumped up and inclined to plunge on, but well enough trained to know that could be a huge mistake. Trail watchers were used by NVA and VC units as an early warning device. Encountering a trail watcher was an indication that an enemy unit was close by and that they would be dug in. With fourteen men, we would be in no position to attack a dug-in force. The platoon, which could only move single file on the narrow finger, had come up. I halted them while I got on the radio to Captain Thomas.

"Charlie Six, this is Charlie One-Zero. We made contact with a trail watcher. I'd like to put some arty up there, over."

"Roger. Wait, out."

After a short wait, Captain Thomas came back on the radio. "Charlie One-Zero, this is Charlie-Six. Negative on the Arty, the range is too close. There are gunships on station. Can you use them, over?"

"That's affirmative, over."

"Roger, they will come down on your frequency and you can control their fire directly."

"WILCO, over."

Charlie-Six, Out."

After another interval, the familiar, warbling voice of a pilot came on the line.

"Charlie One-Zero, this is Snake-eye Three-two, you have targets for me, over?"

"Affirmative, Snake–eye Three-two, suspected NVA position, from my location, direction one-nine-five degrees, three hundred meters, over."

"Roger, One-zero, mark your forward position?"

"WILCO, Three-two, smoke out."

"Gadzinski, throw out a smoke to our front. Any color but green."

Gadzinski tossed a purple smoke grenade out about twenty meters to our front. Then we waited for the pilot's response. In triple canopy jungle, it took time for the smoke to sift through the layers of foliage, and there was always the danger that it would completely dissipate.

"One-zero this is Three-two. I identify Goofy Grape."

"Roger Goofy Grape. Request you fly on a heading of Nine-zero degrees to Two-eight-zero degrees, over."

This was a request that they fly their gun run across our front rather than from rear to front, thereby lessening the possibility of causing casualties from friendly fire. Gadzinski and I were huddled at the base of a huge tree, the roots of which rose out of the ground like the fins of some enormous rocket.

The next thing I knew, the Cobra gunship came roaring in from our right rear with rockets and mini-guns blasting away. It seemed as though we were being enveloped by explosions. Gadzinski and I were tucked into the fetal position, using the gigantic roots for cover. We should have been terrified but instead were preoccupied with communicating with the gunship pilot.

"Negative, negative!" I shouted into the radio, "Fly across our front, over."

"Roger, One-zero."

In spite of me, they proceeded to work over the area as they saw fit. Fortunately, none of our soldiers were hit.

The gunships departed and it was now time to move up the hill, but unforeseen events now conspired to get in the way. First, Bourland appeared saying he had been bitten by a snake. He showed me two puncture holes where the snake had supposedly bitten him. Ever since he had joined the platoon, I had sensed something sneaky and untrustworthy about him. I had a hunch that he was faking in order to get out of the field but I was too cautious to ignore his claim and have him die of snakebite. Doc Wierzalis, our medic, was on the birds that didn't make it to the field, so the medical decision-making was on me. While I tied a tourniquet above the punctures and marked the time, Gadzinski was busy on the radio calling for a MEDEVAC. Because we were in such difficult terrain with no LZ available, medical evacuation would require a helicopter from 95th Evac Hospital which was equipped with a jungle penetrator. This would entail waiting about a half hour until the bird arrived, then the tedious process of lowering a cable with an attached seat through the triple canopy.

The MEDEVAC arrived and we were in the process of guiding the pilot with the penetrator when two or three of my soldiers arrived on the scene, angrily dragging along another soldier.

"We caught this son of a bitch trying to shoot up. He needs to be gotten the fuck outta here."

It was the replacement who had joined us the day before we left. It turned out he was a heroin addict and had freaked out with all the excitement and had tried to shoot up. I was proud of my soldier's intolerance of this and their unwillingness to have him in the field with them. But now what? The simple solution came to me quickly.

"Throw him on the MEDEVAC with Bourland. Gadzinski, call the CP and tell them we're evacuating an additional "pack" and tell them why."

We strapped them on the penetrator seat, sitting face to face and were glad to see them go as the cable was hoisted up through the jungle canopy.

Now, finally, we could resume our movement up the finger.

Totally alert with nerves taut, we cautiously moved forward, scanning our front for the camouflaged outline of a bunker or the subtle firing lanes cut in the underbrush, expecting a burst of fire at any moment.

Reaching the top, we stayed low in the jungle foliage and looked out onto the ridge top, which had been cleared of vegetation. In the middle of the clearing sat a U-shaped bunker, about four feet tall with openings at each end of the U. We watched for a while for any signs of movement. Nothing.

"Get Brown up here with the 203 and send the .45 up."

Jesse "Blooper" Brown was one of the best soldiers in the platoon and a good shot with the grenade launcher. He shortly arrived, bringing with him the .45 caliber pistol that we kept in the platoon. Meanwhile, I stripped off my web gear and helmet and armed myself with the .45 and my machete. I stuck my flashlight in my cargo pocket.

"Put a grenade into each entrance and let's see what happens."

Brown dutifully fired a grenade into each opening. No response.

"OK. Put another one in the right-hand entrance and then cover the opening on the left. I'm going to go in on the right."

As soon as his grenade exploded, I sprinted across the clearing, pistol in right hand, machete in left, and ducked into the bunker. Squatting there alone in the dark, I became more aware of myself. My heart was pounding and my breath was coming in short, ragged pants. I was scared. I was also aware of the coolness, like being in a cave. I listened but heard nothing. Then I felt around for trip wires, again nothing. Reaching into my pocket I got out my flashlight and flipped the switch to 'On' and shone it on the wall to my immediate front.

"AHHHHHHHH..."

There, two feet from my face, the wall was covered with tarantulas — big, hairy, fist-size tarantulas. Recoiling backward, I started wildly flailing at them with my machete. After my initial panic, I calmed down and realized that they weren't attacking me and I could probably just go about my business.

Shining the light ahead, I crawled deeper into the tunnel, probing the floor with my machete as I moved, feeling for trap doors leading to further tunnels below the one I was in. Finding nothing of further interest, I concluded that it was a dry hole. Crawling back to the entrance, I called out, "COMING OUT!"

"OK."

Rejoining the platoon at the edge of the jungle, I sent out patrols to check out the area. The day was almost gone and we would need to establish an NDP in this vicinity. The patrols found some cave complexes and individual caves, but no enemy.

In the NDP that we established, Minh, our useless Kit Carson Scout, was sitting cross-legged near where I was standing with his back to a tree. All of a sudden he exploded into the air and came down hacking at the ground with a machete. Once he got through with his frenzied chopping, we inspected the object of his attention.

"God-damn, look at the size of that centipede."

"It's as big as a lobster!"

Someone had gingerly picked up the segment containing the stingers.

"Look at the size of these stingers. They're a half inch long!"

"Did you see Minh. That son of a bitch has never moved that fast."

"Well shit, I would too if I had one of those things an inch from my dick."

"There it is."

The next day I pushed patrols out further, hoping to find the enemy that I was sure was nearby. Gadzinski received a call from one of the patrols.

"LT. One-two says he has found an NVA super highway."

"Get their location and tell them to hold up until we get there."

Gadzinski and I joined 2d Squad as quickly as we could. Sure enough, they had come upon a very large trail which ran along the ridge top. It was a good ten feet wide and worn smooth with continued use. This had to be an important artery of that vast complex of trails known as the Ho Chi Minh Trail, which brought men and supplies down from North Vietnam.

Using my compass, I shot an azimuth along the trail.

"Gadzinski. Call this in to the Old Man. Give our present location and an azimuth of 100 degrees. Make sure you shack (encode) it up. Tell them we're going to check it out."

While Gadzinski was busy encoding the message and calling it in, I took a closer look around our immediate vicinity. This thing was amazing, one would never know it was there short of coming right up on it. It couldn't be seen from the air because of the jungle canopy.

'These NVA must be like human ants,' I thought.

Once, the report was called in, we continued to patrol the trail to the east, looking carefully for booby traps and signs of enemy presence. We patrolled to the limit of our sector, frustrated again in our search for an elusive enemy.

Landing on Charlie Ridge L-R RTO and SSG Gauthier

Landing on Charlie Ridge, SSG Gauthier on skid

29. RECON RIDGE

"Hey LT," Gadzinski called to me. "The CP just called. They want us to move to a PZ for extraction."

"You're joking! Where do they want us to move to?"

"Give me a minute while I decode the coordinates."

While Gadzinski was de-coding, I started fuming.

"Goddammit, we've only been here for three days and now they are pulling us out? I thought we were supposed to kill dinks. But no, as soon as we get close, they decide to pull us out. This is fucked up."

This is how I, as a young, gung ho and immature lieutenant, looked at it. How the brigade commander might look at it was way outside my field of vision. From his point of view, the last thing he would have wanted this late in the war was a major engagement with lots of casualties. Having established that there were enemy forces in the area, he could pound it with artillery or bombs and not risk the lives of his ground troops and the bad publicity that would have generated.

"LT, they're sending us back to the same LZ where we came in," said Gadzinski.

"OK. Call the squads and tell them to move back to the top of the ridge where we first came up. We'll head back down from there. Call the CP and see if you can find out where we are going."

It was an easy move back to our insertion point. Since we were still only twelve strong, two birds would easily handle extracting us. Meanwhile, Gadzinski had learned our destination, which was Recon Ridge, another mountainous range close to Charlie Ridge.

The LZ where they inserted us was very poor, a barren, rocky, razorback finger covered with little more than waist high scrub brush. Shortly after we arrived, we were visited by a re-supply bird, which carried Bourland and the heroin addict. Bourland seemed to be miraculously recovered from his "snakebite".

One look at the addict told me that he had not had any opportunity to feed his drug habit and was physically in a very bad way. He was going into withdrawal or, in the slang of the day, "The Joneses."

"I'm getting out of here. I don't care what I have to do. I'm getting out of here."

He clearly represented a threat to the safety of the rest of the platoon.

"Give me your rifle."

I took the bolt out of his rifle, put it in my pocket, and handed his rifle back to him.

"Sergeant Gauthier, take his frags and ammo and re-distribute it. You! Stay here with me. I'll see about getting you out of here tomorrow."

His ability to pose a threat to other members of the platoon was now eliminated. As the day faded, his physical condition worsened. By nighttime, he was doubled up in pain, groaning and frothing at the mouth. I had never seen anything remotely like this. My anger toward him was changed into something like pity, that someone could be so enslaved as this. I was also resolved to grant his wish and get him out, no matter what it took. We sat up with him all night, to make sure that he didn't try anything stupid and to do what we could to sooth him—not only for his sake but, more importantly, for the sake of the rest of the platoon.

By the next day, he was such a physical wreck he could barely stand. After being up all night with him, I was drained as well. I got on the radio to the CO and made the case that this guy had to be evacuated, which he agreed to. I was also furious and I vented to Staff Sergeant Gauthier and Gadzinski.

"This is bull-shit, absolute bull-shit! Those ass-holes in the rear only care about playing their numbers game and keeping as many people in the field as they can. To send someone like this yahoo out here is as wrong as two left feet. I'm going to bring it up with Major Lenhart when we get back. This just isn't right."

"There it is, LT. It don't mean nothin'," said Staff Sergeant Gauthier.

One of the battalion log-birds came out around mid-morning and took the heroin addict away, never to return. Our next destination was Hill 502 (so named because of its altitude — 502 feet above sea level), where we would be reunited with the rest of the platoon and get a re-supply. Looking forward to both those events, we moved out, heading ever higher up the ridge.

The day was cloudless and very hot as we trudged up the barren ridge in the scorching sun. Even though we were steadily going uphill, the going was relatively easy. We didn't have to cut trail and the altitude gain was gradual, until we hit the approach to Hill 502. The terrain then rose abruptly for the last hundred meters or so. I had gotten used to humping the mountains by now but this was a killer. Every step was a struggle. Finally at the top, I threw off my rucksack while gasping for breath and watched as the rest of the guys struggled up like overloaded beasts of burden.

Hill 502 had obviously been a firebase at one time. It was completely cleared of all vegetation and ringed with fighting positions. It would be the perfect place to receive our re-supply before moving on. It was a beehive of activity as birds delivered supplies and the balance of the platoon. NCO's were busy breaking down and distributing everything as quickly as possible. Old radio batteries, cardboard boxes, and unwanted rations were put in a hole and set afire, lest Charlie find them and put them to use. As the fires smoldered, we saddled up and moved on, wanting to put as much distance as possible between us and this spot before the end of the day. It was early afternoon and we had a few good hours left before we would have to stop for the night.

Steadily we climbed, ever westward, ever upward. I wanted to get off the ridge and begin patrolling, but it dropped off so sharply on both sides that it seemed impossible to find a place to make our descent. To the north, the vegetation was waist high scrub, but to the south it was dense jungle. In addition to looking for Charlie, we also needed to find water, which was beginning to run short.

Stopping the platoon, I sent each squad in a different direction looking for a spot where we could get down into the draws below us. That's where Charlie would be if he was there and that was where we would find water. A huge tree spanning a ravine looked promising at first but in the end was too dangerous. Sergeant Christiansen approached.

"LT, I found a spot that might work. It's too steep to go down alone, but Bugge has a rappelling rope that we can tie to a tree and with that I think we can make it."

"Outstanding, Chris. Show me."

Sure enough, the spot was as described. It was the best option we had.

"OK, let's do it. The last man is going to have to untie the rope and get down on their own."

"It's Bugge's rope and he is in good shape. I'll have him drop his ruck and we'll get it down some other way. That way he can come down light."

I briefed the other squad leaders and we began the descent. It was like going from day to night in a minute. At the bottom of the draw, under three layers of jungle canopy, it was as dark as late evening. Dark and dank. Leeches were everywhere. They seemed to sense body heat because I could see them start crawling toward me whenever I paused for just a moment.

As we spread out in the draw, a large cave complex was discovered. Again, my excitement rose. Hidden from aerial observation and virtually impervious to artillery or air strikes, this would be the perfect spot for an enemy base camp. It had indeed been used before, as evidenced by bits of equipment and old fighting positions, but not recently. Salazar made the only find of the day.

"Hey, look at this."

Salazar was holding a human skull and, as he lifted it over his head for all to see, its contents fell out onto his bare head.

"Ay chingao!"

This brought plenty of guffaws and a lot of ribbing.

"Hey Salazar, what are you going to do with all those brains?"

"Salazar's the brainiest guy in the platoon. Maybe he should be in charge."

"Aw, fuck you guys."

"Hey, it don't mean nothin'."

"Yeah, there it is."

Even though there was no sign of Charlie, we were able to find the much needed water.

The next day found us on the move again through the roughest, thickest terrain we had seen yet. The vegetation was like a solid wall and it was necessary to cut trail every step of the way. Our trail looked like a narrow hallway in an endless green, leafy building. Sometimes the going got so tough that we had to backtrack and find another way. It was re-supply day and we hadn't found an LZ or even a spot where the helicopter could drop our supplies from the air. By early afternoon it became clear that no LZ was going to materialize on its own. Stopping the platoon, I summoned Staff sergeant Gauthier.

"We are going to have to cut our own LZ. That area over there is pretty flat and it's big enough for a Huey to get into. Give each squad a section and let's start cutting."

"Roger that, LT."

Every soldier carried a machete and everyone, including myself, set to work clearing an LZ. This involved chopping down some pretty large trees in addition to all the jungle undergrowth. In a couple of hours, we had cleared a patch of jungle big enough for a helicopter to land in. By the time we finished, each of us was drenched in sweat as if we had jumped into a lake fully clothed.

When the re-supply arrived, off jumped a new replacement named Smedlap and Thorkelson, an old-timer who was returning from a few days in the rear. In addition to our normal supplies we received, of all things, a hot meal of fried chicken, the first and last I was ever to have in the field. We named the LZ 'LZ Chicken'. I greeted Smedlap, the replacement. His only comment to me was, "This doesn't look so bad."

This struck me as a very odd thing to say in the middle of the jungle, surrounded by men who were exhausted and soaked through with sweat. In a few days, he would be eating those words.

Next I talked to Thorkelson. "Welcome back Thork. How you doin'?"

"OK, LT, except I heard that Frenchie got killed."

"No shit? What happened?"

"He was with B Company over in Happy Valley and got blown up by a booby trap."

"Man, that's fucked up. He was a good guy."

Frenchie was an artillery recon sergeant who accompanied the infantry as a Forward Observer (FO) to help coordinate artillery support. He was outgoing and personable and his modest rank of buck sergeant didn't seem to affect his outlook on the world one way or the other. He was as comfortable around officers as he was around junior ranking enlisted men. When we were in the base camp he was a frequent nocturnal visitor to the lieutenant's hooch to chew the fat.

Author

SGT Ricardo "Rod" Rodriguez

Author

Moving down mountain stream

30. PUSHING THE BOUNDARIES

We continued to patrol the Recon Ridge area. I mainly kept the patrols in the streams. Everyone has to have water and I surmised that we would find enemy base camps on or near the sources of water. Also, it was easier to move quietly in streams. I further came to the conclusion that the VC/NVA were staying just beyond the range of our artillery and, hence, our boundaries so I led patrols right to the limits of our area of operations. Still, nothing. As the days and nights dragged on, an idea took hold in the platoon.

"Hey, when we get back to the rear, we should have a platoon party."

"Yeah, no shit. We could go down to Gunfighter Village to one of the EM clubs and have it there."

"Whaddaya think, LT?"

"I think that's an outstanding idea."

"Would you come?"

"Yeah, I'd definitely stop by."

In order to maintain good order and discipline, officers were not supposed to fraternize with enlisted men. The extreme adherents to this dictum shunned all association with their men outside of official contact, while others, perhaps most, would socialize with their men to one degree or another. I fell in this group. After all, I lived with these guys twenty-four hours a day, seven days a week, month in and month out and I asked much of them. I felt it would be the height of disrespect to shun them when we were in the rear.

On the 29th of March we were extracted from the field for a five-day stand down. After weapon cleaning and inspections, planning the party was uppermost in the minds of my men. I had other fish to fry. I was still seething about the heroin addict and I couldn't wait to confront Major Lenhart with his malfeasance. Spying him outside in the base camp, I approached him and saluted.

"Sir, Lieutenant Sherwood. I have a complaint to make."

I recounted the incident and concluded by saying, "Sending these guys to the field jeopardizes the men and the mission. It isn't right."

What was I thinking? In a profession that runs on overdoses of testosterone and machismo, what did I think I was going to accomplish by confronting "The Bruiser" in public. He glared at me from under bushy eyebrows in his bulldog face and delivered the worst insult that an infantry officer could receive. "Lieutenant, you have a leadership problem."

Then he stalked off, leaving me standing there, blushing with anger and impotent rage. I had disliked him since my first night in the unit but this sealed the deal. Someday, somehow, I would get even.

On the day of the platoon party, my buddy John Beasley and I went down to the Bamboo Room restaurant in Gunfighter Village. My plan was to have a nice meal with Beas and then make an appearance at the party. Over shrimp fried rice and champagne we engaged in the usual discussion about when the brigade would stand down. The big NVA offensive to the north of us was also of concern.

"So when do you think we'll stand down?"

"Fuck if I know. But it probably won't be long."

"You heard that One-One Cav already stood down?"[7]

---

7. 1st Squadron, 1st Cavalry Regiment, the Brigade's cavalry squadron.

"Yeh, and I understand our own A Company is going to stand down any day now."

"You know what's got me worried? While all this standing down is going on, the NVA offensive on Hue and Quang Tri is just up the road. What if the ARVN's can't stop them? What if they want us to go up and help? We could be in deep shit. They've got tanks and everything."

"I'm going to start carrying LAW's[8] with me when we go back to the field."

"That's a good idea. I think I will also."

"Well, I'd better mosey on down to the EM club. My platoon is having a party and I promised I'd come by."

"OK. I'm going to head on back to camp."

"Alright, I'll see you a little later."

The EM club was only a short distance away. When I arrived, the party was in full swing. The "club" was nothing more than a big rectangular room with a bar and tables and chairs. I grabbed a beer and sat down with Sergeants Rodriguez and Christiansen.

"How's it goin'?"

"Real good LT, glad you could make it." I had barely gotten into the small talk when there was a gigantic eruption to my left on the other side of the room. Glasses were crashing on the floor, chairs and tables were knocked over and a full on fist-fight was taking place.

Rod, Chris, and I, ran over and started breaking it up. I grabbed one and hustled him out the back door while calling to the two sergeants to get the other fighter out and send him back to camp. The pugilist I had charge of was Bourland, the soldier who I suspected had faked a snake bite in order to get off Charlie Ridge. Did the fight have something to do with that? I never found out. Bourland's lower lip was split in two

---

8. LAW—Light Anti-tank Weapon—a portable, disposable bazooka type weapon.

and blood was dripping down his chin. He was very agitated, panting and pacing.

"I'm going to kill him," he said.

"Bull shit you are. It was man to man and you're gong to take your ass-whipping like a man."

I kept Bourland for a suitable interval until he had calmed down and his opponent had plenty of time to clear off.

"Get back to camp. I'll deal with this when I get back."

I went back into the club, where the party had resumed, less the two fighters. Or so I hoped. The next thing I knew, two other soldiers got into an argument about the original fight and started fighting. After breaking up that fight, I declared the party to be over.

"Alright! The party's over! Everyone back to camp."

I stayed behind to make sure everyone moved out. There were enough sergeants present to make sure everyone kept moving and kept the peace. When everyone seemed to be gone, I headed back toward camp. When I got to the gate, I had a feeling that I should go back for one last check to make sure that no one was left behind. Sure enough, when I got over in the vicinity of the EM club, there was an Air Police jeep with Warner and Bugge sitting in the back. They had nothing to do with either fight and there was no reason why they should be detained.

"Why are these men under arrest?" I demanded.

"They aren't under arrest, sir."

"Bull shit. You're preventing them from going about their business, therefore they are under arrest."

"They aren't under arrest, sir, they're just being held."

"Well, unless you are going to charge them with something you have no reason to hold them. They are my men and I'll take responsibility for them."

"Alright, sir, I'll sign them over to you."

I signed the necessary papers and left with Warner and Bugge in tow. Near the gate, we ran into PFC Eddie Sosa, who joined us. Out on the road, a captain stopped his jeep to give us a lift back to Camp Crescenz. All four of us crammed into the back of the jeep, which was a very tight fit.

"What unit are you in?" The captain asked.

"First of the Forty-sixth." I replied.

At that point, Sosa, who was drunk, weighed in.

"We're grunts, sir! Grunts to the max. We go for the kill. GO FOR THE KILL!"

To add emphasis, Sosa was slapping the captain on the back with both hands. The captain was a rear-echelon type and seemed rather timid. He didn't know what to make of this and I'm sure he was glad when he dropped us off at the camp gate.

I immediately proceeded to the company headquarters to report the incident to Captain Thomas.

"Sir, my platoon was having a party down at the air base and there was a fight. It was not a real big deal and I'd like to handle it within the platoon."

"So what happened?"

I explained in more detail what had transpired.

"OK. Go ahead and handle it your way."

About an hour later I was summoned back to the orderly room to see Captain Thomas.

"We have to go see the brigade commander. Now."

We got in his jeep and made the drive to brigade headquarters. This obviously had to do with the events at the air base and I was concerned, but not overly so. Fights happen, especially among men who spend weeks on end wound up with fear and tension and are suddenly unwound by the effects of alcohol. I figured the brigade commander would ask me what had happened, I would explain, and he would tell me not to let it happen again.

At the headquarters, we went through the ritual of entering the office of a senior commander. Three knocks on the open door.

"Enter."

Marching to a spot two paces in front of the desk, I executed a sharp left turn and saluted.

"Sir, Lieutenant Sherwood reporting as ordered."

Brigadier General Joseph McDonough returned my salute but did not tell me to stand at ease so I remained at attention. This was my first encounter with the brigade commander. Thinning reddish blond hair that was beginning to grey fringed a mostly bald head. He had a broad, Irish face, fair skin, and a very stern countenance. I could tell from the badges on his uniform that he had seen combat service in Korea and Vietnam.

"It has been reported to me that you were leading your platoon through the Da Nang Air Base tearing up clubs and beating people up."

What the hell? Where did that come from? I thought, *That isn't what happened at all. I need to explain what really happened.*

"Do you understand your rights under Article 31 of the Uniform Code of Military Justice?"

Holy shit! I thought, *He's reading me my rights. This is serious!*

"Yes, sir." I heard myself croaking.

"You have the right to remain silent. Anything you say or do can and will be used against you at a trial by court martial …." And on through the familiar recitation of the Army's version of our 5th Amendment rights.

I was stunned. I barely heard him as he concluded and asked me again if I understood. Now was the time to explain to him what had really happened.

"Yes, sir. Permission to speak, sir."

"Denied. You're dismissed."

Thoroughly shaken, I saluted, executed a left face and marched from his office.

Captain Thomas said, "Well, it looks like you won't be handling this in the platoon after all."

"What do you think is going to happen?"

"I don't know. We'll just have to wait and see."

By the next day, things were not looking good. Three of us were being charged: myself, Bugge and Warner and there was talk of giving us an Article 15, which is a form of non-judicial punishment which becomes a permanent part of one's record. For an officer, that could be a career stopper. I was a hot stew of mixed emotions. I was angry that such a big deal was being made over something that I saw as trivial. Bugge and Warner hadn't done anything wrong and the worst thing I could be accused of was fraternization with enlisted men. I was also sad and frightened that my chosen career could come to an ignominious end before it even got started.

"I won't accept an Article 15. I'll demand a trial by court martial. What do I have to lose? If I take an Article 15, I'll be screwed. At least with a court martial I would have a chance."[9]

My men were solicitous, but, since none of them aspired to an Army career, they tended to see things from a completely different perspective.

"Hey LT, don't sweat the small shit. I mean, what are they going to do to you, send you to Vietnam?"

"Hah, there it is."

In the end, it was decided to give me a Letter of Reprimand. I was marched before Lieutenant Colonel Perkins, the battalion commander. Somewhat to my surprise, Captain Thomas spoke up for me and my

---

9. Because an Article 15 is non-judicial, a soldier has the right to refuse it and demand trial by court martial, which is a judicial proceeding.

men. After he finished, Colonel Perkins turned his attention back to me.

"Do you have anything you want to say?"

"Yes sir."

After reviewing the facts of what had actually happened, in contrast to the wildly exaggerated report that the brigade commander had received, I went on to say, "Day after day we hump the bush, never knowing what's going to be around the next bend in the trail, and the men get wound up like a steel spring. When they let off steam, sometimes they come uncoiled."

"We haven't been in that much contact, so there's no reason for the men to be so wound up, as you say."

This was rich, coming from a man who had not served as an infantryman in Vietnam prior to this assignment and, to my knowledge, had never set foot in the field. He had no idea what it was like out there.

Having had my say, Colonel Perkins continued, "I've decided to give you a Letter of Reprimand, which I will read to you now. 'You are hereby reprimanded for conduct unbecoming an officer blah, blah, blah, will not be tolerated, blah, blah, blah, this letter will be made a part of your personnel record.' Do you have any questions?"

"No sir."

After receiving my copy of the letter, I saluted and marched out of his office.

As soon as I got back to my hooch, I crumpled the letter up and threw it away. This was not an act of anger or defiance but one of deep shame. I had tried so hard to be a good officer and now I had this blot on my name, on my honor. This was a curious reaction from someone who had been a regular fixture in the principal's office since first grade and who wore as a badge of honor the number of times he had received corporal punishment from the dean while in secondary school. I guess the difference was that, unlike school, which was forced on me, I had chosen this profession.

Staff sergeant Gauthier sensed my angst. "Don't sweat it, LT. I have a friend who works in personnel. I'll get him to pull it out of your file."

"Thanks, Sergeant Gauthier."

He was as good as his word. Shortly before rotating back to the States, he presented me with the copy from my file. It, too, I wadded up and threw away. This episode was something I was deeply ashamed of for a long time. Now, looking back, I wish I had kept the letter as a keepsake to have a good laugh over.

31. SOME LIKE IT HOT, SOME LIKE IT COLD

On 3 April, we were assembled on the pad waiting for the birds to take us back to the field. The air base episode was concluded and even though I didn't like how it had turned out, my mind was clear and focused on the task at hand. The lead platoon had already lifted off and my platoon was next in line. The day was grey and overcast but was considerably brightened by the presence of about four "Donut Dollies" (Red Cross volunteers) who were on hand to give us cheer. These were the first American girls I had seen since leaving the States.

"Whoa look, round-eye chicks!"

"Son of a bitch! I ain't seen a round eye in seven months!"

Everyone was engaging in an amazing blend of macho bravado and light-hearted, flirtatious banter. The girls seemed to be eating up the attention as well. We felt like knights entering the tournament, all vying for the favor of a fair maiden. Jim Cole, the company executive officer, was orchestrating the lift from the edge of the pad. In the midst of our revelry he called out, "THE LZ IS HOT!" (meaning that the LZ was receiving enemy fire)

Suddenly everything went silent, like someone had pulled the plug on the jukebox or as if, in the middle of Christmas dinner, someone announced the death of a family member. Everyone went quiet and grim. I immediately thought of Tony Zanotelli, my good friend from high school, who landed on a hot LZ and was shot before he could even get off the chopper. I yelled, "LOCK AND LOAD!"

A unit was most vulnerable when it was landing, and a hot LZ was the worst possible scenario. I was scared and I knew everyone else was but, looking around, all I saw were grim, resolute faces. No one wavered.

Our lift arrived shortly. According to plan, I jumped on the port side of the lead aircraft, along with Gadzinski.

After flying west for several minutes in the direction of the mountains, I could see the LZ in the distance. It was a large, grassy field near the base of the mountains. To the north there was a low hill overlooking the LZ. Was that where the enemy was? Or were they right on the LZ?

As we began our final approach, the door gunners opened up with their pedestal-mounted M-60 machine guns, spraying the ground on either side of the LZ, as well as the tree line on the low hill. As the gun on my side traversed to the left, its muzzle was right next to my right ear. The noise was such that I thought my head was going to explode.

Now I could see yellow smoke marking the spot for the lead bird to land. The rotors flared in preparation for landing, plastering the grass to the ground. We were about six feet off the ground and the crew chief started yelling and motioning with his arms.

"GO, GO!"

All eight of us leaped into space, weapon in hand and rucksack on our backs. No sooner were we in the air than the birds were gone.

I landed in a bomb crater that was filled with water, which absorbed the shock of landing but left me floundering like a fish left on the shore when the waves recede. Collecting myself I crawled out of the crater. I could see nothing. In the absence of the rotor wash, the elephant grass had returned to its full seven-foot height. Visibility was about one foot.

I couldn't hear any firing but I kept low to the ground just in case. My first task was to get on the radio and find out what the situation was.

"GADZINSKI!"

"OVER HERE!"

Crawling through the grass in the direction of Gadzinski's voice, I found him about ten feet away. We raised Captain Thomas on the radio.

"Roger, six, what is your location relative to twelve o'clock, over?"

"Six, here. I'm at twelve o'clock, over."

That made sense, since he would have been on the lead bird of the first lift. And, since I was on the lead bird of my lift, he couldn't be far away.

"Roger, I'm moving to your location, out."

Crawling in the direction of where I had seen the smoke grenade, I soon found the CO.

"So, what's the situation, sir?"

"One of the ships took fire as we were coming in. It must have been a sniper. There hasn't been any more firing so he must have di di mau'd. They may still be around so we'll need to search the area. From here I want you to search from twelve to four. Second and third platoons will search from four to eight and eight to twelve. Go out about a click."

"Roger that, sir."

He showed me on the map what the boundaries would be on the ground between me and the other platoons.

Because of the zero visibility, it took a while to find the squad leaders and for them to find their men. Because of the possibility that the enemy were still around, everyone was cautious and keeping low to the ground.

I broke the platoon sector into squad sectors and instructed the squad leaders to patrol their sectors using the cloverleaf pattern we typically used.

At around 1500 hours, after much fruitless beating of the bush, I got a call from Sergeant Merrick, the 3d Squad Leader. "One-zero, this is One-three. We've got some tunnels with fresh tracks around them, over."

"Roger, One-three. Give me your location. I'll be there ASAP, over."

"WILCO."

After receiving and hastily de-coding Sergeant Merrick's location, the squad I was with moved out and arrived in about 20 minutes.

"We found these tunnels, LT. They're all empty but these tracks around them are fresh."

I squatted down and examined the tracks, which had clearly been made by Ho Chi Minh sandals[10].

"They're fresh all right. They're still wet and the edges of the imprints are sharp. These are probably the same guys who fired at the birds on the first lift. They obviously hatted up once we were on the ground. What direction do they take?"

Gesturing west toward the mountains, Sergeant Merrick said, "We followed them in that direction for a ways, but then they disappear when the ground gets hard."

"OK. We'll be heading up there anyway, maybe we'll catch them. Good work."

"Thank you, sir."

---

10. Ho Chi Minh sandals was the name given by the Americans to the footwear typically worn by the VC and many NVA. The sole was made of tire tread and the straps were cut from inner tubes. They were very serviceable.

32. WHAT ARE YOU GOING TO DO NOW, LIEUTENANT?

The next day we were heading west, back up into the mountains. Just before the ascent got steep, I stopped the platoon for a five-minute breather. As usual, almost everyone, including myself, lit up a cigarette. Gadzinski and I flopped down and eased out of our rucks.

"Well, LT, here we go again, more mountains."

I took a deep drag and exhaled slowly.

"Yeah, they're a bitch all right, but I'll take these any day over working villages where you don't know who is who. At least out here it's a free fire zone. If you run into anyone, you know it's Charlie."

"There it is."

As I got to the end of my smoke, I twirled my finger in the air and said, "Alright, saddle up."

As we struggled into our gear, Sergeant Stringfellow came up to me. "LT, Smedlap (pseudonym) says he isn't moving."

"Whaddayou mean, he's not moving!?"

"There it is, sir, he just said he wasn't going any further."

I was hacked off. Smedlap was a recent replacement who had commented upon his arrival in the field that it didn't look so hard. He had done nothing to fit into the platoon and was looked on as something of an outsider. Sergeant Stringfellow was not the hard-boiled type of NCO who would handle problems like this physically without bringing it to the attention of his officer.

I went down to where Smedlap was sitting by the trail with his helmet and rucksack off. He had sandy brown hair, a sallow complexion, and hooded eyes that revealed no emotion. Indeed, he displayed no emotion, not anger, fear, sadness — nothing.

"What's the problem?"

"It's too hard. I don't see why I should have to do this. I'm not going."

"What do you mean, 'it's too hard'? You're the one who said that it looked so easy. I'm not asking you to do anything that I or anyone else isn't doing. So get your ass in gear and let's go."

"No. I'm not going."

Most people who have not been in the military imagine it as a place where authority and discipline are absolute, where all orders are robotically obeyed. But the fact is, the whole system relies on the general acceptance of authority, rules, and regulations, just as in civil society. In war, the responsibility for ensuring that orders are obeyed rests with the leader on the spot. There is, on the spot, no higher authority that can force compliance.

I found myself in one of those, 'What are you going to do now?' scenarios like we had encountered in our officer training — an unexpected situation where there is no clear-cut solution. Whatever solution gets the job done is a good solution.

"OK. See you later. SADDLE UP, LET'S GO."

I resumed my place near the front of the platoon and we moved out. No one seemed concerned about leaving Smedlap behind. Perhaps they viewed him as a hindrance that would not be missed.

Now it was a contest to see who was going to blink first. Abandoning a soldier in the field was a very serious move on my part but I was not about to let this guy call the tune or hold us up. The trail we were following bent to the left as it climbed a ridge. As I made the turn to the left, I glanced surreptitiously over my left shoulder to see what Smedlap was doing. I was very uneasy about my decision, but resolute.

After we had gone about one hundred-fifty meters, Smedlap got up and started following us. I halted the platoon long enough for him to catch up and then we continued on our way. Relieved, I lit up another cigarette.

Later, when we were in the rear area, I checked Smedlap's personal file in the company headquarters. His general intelligence score was in the 70's and some of his aptitude scores were in the 40's. The average infantryman had scores in the 90's or low one hundreds, which indicated good but not extraordinary intelligence. Although his record did not specifically identify him as such, Smedlap was undoubtedly one of Secretary of Defense Robert MacNamara's Hundred Thousand, a project which drastically lowered the military entrance requirements in order to meet rising manpower demands. This and other policies initiated by Secretary McNamara have left him indelibly etched in my mind as one of the arch villains of that war.

33. NVA EASTER OFFENSIVE

On 30 March, the NVA launched a massive, three-pronged offensive. The northern prong was aimed at Quang Tri and Hue. By late April, the NVA forces had pushed very close to Hue, only 60 miles north of us. This was a full-on assault by conventional forces consisting of infantry, tanks and artillery. We had a low opinion of the capabilities of the South Vietnamese Army, referred to as ARVN, and were extremely concerned about what their success or failure might mean for us.

"Hey LT, it sounds like the shit has really hit the fan up at Quang Tri."

"Yeah, I sure hope the ARVN's can hold them."

"Whaddaya think's going to happen, LT? Think we'll have to go up there?"

"Hell if I know. I'm afraid if the ARVN's break, they will send us up and we don't have any tanks or heavy anti-tank weapons. Even if they didn't send us up, there would be nothing between the dinks and us. Either way, our asses would be out."

"Shit! Marvin the ARVN can't fight his way out of a wet paper bag. You know that, LT. Charlie'll kick his ass."

In the event, some ARVN units fought valiantly and some didn't, but the weight of the offensive was more than they could bear. President Nixon ordered massive B-52 strikes, code named Arc Lights, to support the ARVN forces and it was this that ultimately stopped the NVA offensive. The strikes could be felt all the way down where we were. The ground would tremble and a distant rumble could be heard.

"God damn, what is that?"

"That's an Arc Light. They must be putting one in on the dinks up north."

"Whoa, that's going to put Charlie in a world of hurt."

"There it is."

We continued our mission in the mountains, expecting at any moment to be told to assemble at some PZ to be airlifted to the north. We had armed ourselves with LAW's (Light Anti-tank Weapon) but these would have been pretty feeble against the kind of onslaught we would have faced. It was with relief some weeks later that we learned that the offensive had been stopped.

34. JUNGLE SORES & FRIENDLY FIRE

As we climbed higher and higher into the mountains, the elements were as much an enemy as the VC and NVA. The monsoon season was over but Mother Nature seemed to be experiencing drastic mood swings. At one moment, the weather would be blazing hot which, with the high humidity, made it seem like breathing through a wet washcloth. Then, suddenly, it would cloud up and dump torrential rain on us. We were wet all the time, either from sweat or rain.

The constant dampness probably contributed to the jungle sores which plagued many of the men. These were a serious health hazard, usually requiring evacuation after the medic had performed the preliminary first aid. One afternoon, after stopping for the day, Watkins reported to Doc Kroze with a jungle sore on his head, above his left ear. Sores typically appeared on the arms, legs, or back. This was the first time any of us had seen one on the head so we gathered around to watch the treatment. Watkins crouched down to give Doc easy access to the wound, gripping the trunk of a bush for stability and in preparation for coping with the pain which was soon to come.

Doc began by squeezing the sore to work the core out. This was the most difficult, time-consuming, and painful part of the process. Watkins' teeth were clenched as he channeled his pain into his grip on the bush, his knuckles white with the effort. As the core slowly emerged, it was the usual semi-solid, whitish substance reminiscent of congealed bacon fat.

"Goddamn, look at the size of the core."

"Fuck! It's as big as my little finger!"

"No wonder you're so fuckin' dumb, Watkins. That shit ate your brains up."

"Fuck you guys."

"Aw, come on man, we're just playin' with you. It don't mean nothin'."

"Yeh, there it is."

Once the core was completely out, Doc started filling the hole with sterile packing that looked like white cotton string.

"How far in does that hole go?"

"Doc! You think you got enough string?"

After packing was completed, the wound was smeared over with bacitracin ointment and covered with a field dressing.

"OK Watkins, that will hold you for now. Make sure and keep it dry. We'll get you evacuated ASAP."

A couple of nights later, we set up for the night on a narrow ridge. This spot had been occupied in the past. It was cleared of vegetation and had deep fighting positions dug, the bottoms of which were layered with empty C-ration cans left behind by previous occupants.

I plotted and called in our night defensive artillery targets and had mechanical ambushes placed on the only two approaches to our position. After checking the perimeter, I settled down to my usual dinner of C-ration Beef with Spiced Sauce over ramen noodles. I had been in country for four months now and had lost about thirty pounds. I knew nothing about nutrition and didn't realize that I was slowly starving myself by consuming far fewer calories than I was burning.

All was quiet until around 2100 hrs when another unit's artillery illumination mission just to our south started showering us with shrapnel from the illumination canisters.[11]

"INCOMING," I yelled, as I grabbed the radio and started calling the company CP.

"Charlie six, this is Charlie one zero, we are receiving incoming from a friendly illum mission to our Sierra. Check fire, Check fire, over."

No response.

"Gadzinski, keep calling the CP, I'm going to make sure everyone is under cover."

As I jumped up to run down the ridge to Sergeant Rodriguez, I saw Salazar make a racing dive into one of the fighting positions. Over the noise of the artillery, I could hear the clatter of cans as he hit the bottom.

"Rod. Get around the perimeter and make sure everyone is under cover while Gadzinski and I try to shut this off."

"Roger that."

I ran back up to Gadzinski. "I got through to the CP", he said. "They're calling for a check fire."

This meant notifying the artillery forward observer who was part of the CP. He, in turn, would have to contact the firing battery and call for a check fire. In the meantime, the mission continued and we continued to be showered with debris whistling down and landing all around us. Although it seemed like much longer, it was over in a matter of minutes. It should have been terrifying and maybe it was for the soldiers who had nothing to do but seek cover. But for the NCOs, Gadzinski and myself, we were too busy to be frightened.

---

11. When an illumination round burst in the air, it discharged a canister, which sent debris raining down much like shrapnel.

It was over as quickly as it started and the jungle returned to its own set of night noises. Soldiers crept from holes and talked quietly about the excitement.

"Man, that shrap-metal was going everywhere."

"Did you see Salazar diving into that hole?"

"Yeh, man, that was a trip."

"A piece of that scrapnel landed right next to me. A few inches closer and my ass would have been out."

***

After a few more days of fruitless patrolling, we were abruptly ordered on the 9th of April to find a PZ and prepare for extraction. This was only day seven of what was supposed to have been a twenty-thirty day mission. This was alarming news because it seemed to confirm our fears that we would be sent up to Hue or Quang Tri in support of the ARVN's, who were reportedly having a very tough go against the NVA onslaught.

35. UNCERTAIN FUTURE

Back at Camp Crecenscz, many contingency plans were being discussed regarding the battle to our north. Among the soldiers, the feelings were as varied as the soldiers themselves. For myself, my feelings oscillated between a brash desire to finally lock horns with an enemy that had so far completely eluded us, to a more realistic assessment that engaging in this battle would have a bad outcome and accomplish nothing.

In the end, we were not committed. I'm sure that somewhere in the Army's archives there is a record of the decision-making process that resulted in us not being sent north against the NVA. My guess is that the decision was made at the highest echelons and that the reasoning went something like this:

1. We were a light infantry unit, without armor or anti-tank weapons, making us relatively ineffective against a conventional force consisting of infantry, armor, and artillery.
2. As a brigade, our contribution would be minor against a multi-divisional assault force.
3. Our commitment would likely result in many casualties, something unacceptable to the American public in 1972.
4. Our involvement and subsequent casualties could trigger

the commitment of yet more U.S. ground forces, also unacceptable to the American public.

The bottom line was that the U.S was withdrawing from Vietnam. Vietnamization of the ground war was the name of the game and the ARVN would have to go it alone.

But I had more personal things to worry about. Captain Thomas announced to me, "Lieutenant Cole is being moved to the S-1[12] shop and I want you to be the XO."

The company Executive Officer is the second in command of the company and is, by the table of organization, the senior lieutenant in the company. He is responsible for overseeing the administration and logistics of the company, leaving the commander free to maneuver the company in the field. Some might have viewed this as a promotion and been pleased. Some might have been pleased with the opportunity to get out of the field. I was neither. I was horrified. The thought of being stuck in the base camp with all the other rear echelon types was more than I could bear. Field duty was hard but it was the essence of my chosen profession.

"WHAT!? I don't want to be the XO!"

"I don't care what you want. I want you to be the XO."

"The XO is the senior lieutenant in the company. I'm not the senior lieutenant. Tom Mylan has date of rank on me and he has been in-country longer."

The argument didn't end here but I had blunted the attack by falling back on Army custom and protocol as my defense. For the

---

12. The S-1 is the battalion staff officer responsible for personnel and administration. Jim Cole was going to be the Assistant S-1)

moment, I was left dangling on the hook of uncertainty about our deployment to the battle in the north, and on my future as the leader of 1st platoon.

Both battles raged on, with me sticking to the only defense I had, which was that Army customs and protocols must be adhered to. These are not ironclad and commanders may override them as they see fit, but I had no other basis for argument.

Within two days, both matters were resolved. We were being re-deployed to the field in our regular AO, and I was still the platoon leader of 1st platoon. Tom Mylan was now the XO. I was never made privy to why I won my case. All I cared about was that I won it.

On the morning of 11 April, we were assembled on the chopper pad in dense fog, waiting to be flown back to the field. Our lift helicopters were socked in at their base on Marble Mountain. As the morning wore on, the fog remained and so did we. There was some speculation that the mission might be postponed until the next day, but the battalion seemed determined that we were going to get to the field.

In the early afternoon, word came down that we would be ferried to the field by Chinook. The Chinook was a large dual rotor helicopter that could hold the entire platoon. For some reason, soldiers dreaded riding in a Chinook, possibly because, when they crashed, a lot of people were killed.

I began to hear the now familiar oaths and declarations of non-compliance. I had grown used to it and no longer paid it much mind.

"I ain't ridin' on no shit-hook."

"No sir, buddy. Those things are dangerous."

As the Chinook arrived with a mighty roar, everyone scrambled for cover from the very real danger that the tremendous rotor wash would pry up the large steel plates that comprised the pad and send them sailing through the air like playing cards.

After settling on the pad, the tail ramp lowered.

"LET'S GO! LET'S GO!"

Everyone emerged from their cover, one hand holding their helmet down, the other holding their weapon, and scrambled aboard, settling into the webbed nylon seats that lined the bulkhead. Once everyone was aboard, the ramp closed and we were encased in the belly of this large, flying beast. The pilot increased the pitch of the massive rotors and we slowly lifted off and were airborne. There were a couple of porthole windows but, for the most part, we couldn't see where we were going.

After flying for several minutes, I could tell we were approaching our destination because the aircraft began to slow and the noise from the rotors changed to a slower, more deliberate rhythm. The tail ramp began to lower, allowing sunlight to stream in. Once it was fully lowered, I looked out the back to see…the top of a mountain-top firebase!!

"What the hell are we doing here?" I wondered.

We were hovering over the fire base helipad, which was too small for the Chinook to land on. We had to jump from the ramp onto the helipad. Just as Bugge was getting ready to jump, the Chinook bucked, rocketing him into the air as if he were jumping on a trampoline. He landed in the mass of concertina barbed wire that surrounded the base, his steel pot bouncing off down the hill into a further tangle of concertina. Several soldiers helped extricate him from the wire. He was pretty cut up but otherwise uninjured. He seemed pretty pleased not to have to wear a helmet and immediately produced from his rucksack an Australian-style bush hat to cover his head.

We were told that the reason for landing on the firebase was that there was no suitable LZ for a Chinook in the mountainous jungle. I was given grid coordinates for our destination, which necessitated walking down the mountain and then up onto a distant ridge. We negotiated a zig-zag lane through the defensive wire and were on our way. It took us three days to reach our original destination. One of the soldiers summed it up pretty well. "Ain't this some shit? It takes us three fuckin' days to hump to a place that they could have flown us to in twenty minutes."

Ours was not to reason why.

Los Cuatro Amigos Front-- SP4 Danny Salazar, SP4 Roberto Granado, Rear--PFC Garcia and SGT Rodriguez

36. RUMORS, REPLACEMENTS AND A BATH

After our drop-off at Fire Support Base Linda on 11 April, we humped steadily up into the mountains for three days before reaching the spot where we were originally to have been inserted. There was an LZ there so we received a re-supply. Much to our disappointment, the orange nylon bag containing the mail was absent. Letters were as precious as gold and none of us had received a letter in quite a while. The only consolation was that there seemed to be a SNAFU in the mail system and we were all in the same boat.

After getting re-supplied, we plunged into the dank, leech-ridden jungle off the south side of the ridge. Leeches were always a feature of life in Vietnam but this area had an abundance that we hadn't seen before. Every time we stopped, you could see them inching along the jungle floor in our direction. In many ways, the leeches were like the VC/NVA—always present but seldom seen, leaving the occasional reminder that they had been there. Checking myself one morning, I noted that at least three of them had nourished themselves on my blood. Where once they had been a source of revulsion, leeches were now just a routine part of life.

Through the Stars and Stripes newspaper and other news sources, we had been hearing of all the anti-war protests back home. The most recent protests seemed to be about the bombing campaign in support of the ongoing ARVN defense against the NVA offensive. Because there was nothing between us and the NVA but the ARVN and the bombing campaign, we took these protests very personally. One evening, after settling in, Sergeant Rodriguez commented on one such article. "LT! Did you see the article in the Stars and Stripes about the protests back home?"

"Yeh, that really gives me a case of the red ass. I'd like to get those sons of bitches over here and see how long they sympathize with the NVA."

"No shit. It's one thing to want us home, but to be in favor of the vatos that are trying to kill us, that pisses me off."

Even with stand down imminent, we continued to get replacements, some fresh from stateside and others from in country. In addition to spreading the latest rumors, they also brought news from the rear. Word reached us that First Sergeant Rollins had been held over his rotation date because he was considered mission essential. Also, our sister battalion, 2d Battalion, 1st Infantry, got some bad press in the stateside newspapers.

"Did you hear what happened to Top? He got held over his DEROS."

"Oooh! Are you kidding? You know The Bear has the ass over that."

"That's pretty chicken shit if you ask me."

"Hey, did you hear about Two-One?"

"No, what?"

"Well, you know they got sent up to Phu Bai?"

"Yeh."

"Well, when they got there, they were told to cross this area that was supposed to be heavily mined and they supposedly refused to go. Before you know it, it's on the front pages of the papers back in the world."

"No shit?"

"Yeh, and the bitch of it is that they went ahead and went across."

"I hate these fuckin' reporters. They're just looking for anything negative to say about us."

"There it is. They've already got their story, all they have to do is find names and faces to fill it in."

Day after day we continued patrolling: along ridges in some of the highest mountains we had seen, then frequently plunging down into the permanent twilight of the jungle forest where we followed mountain streams which cascaded down from the highest points. This was ideal terrain for the enemy to hide in but we found no sign of him whatsoever.

We knew that when this mission was over, the company was to go on support duty for two weeks. We knew what two of the three assignments were: one platoon would guard 24 Corps Headquarters and one platoon would guard the 95th Evacuation Hospital. We didn't know what the third assignment would be but we figured that it would be equally easy duty like the other two. The soldiers increasingly talked in eager anticipation of this period of relative luxury.

"Which one you think we'll get?"

"Fuck if I know, but I can't wait."

"Yeh, clean beds, three hots a day, no humping."

"I hope we get 95th Evac. They got all them nurses there."

"Oh yeh, Romeo, I'm sure they're just waiting for you to get there."

"Fuck you."

"Hey, I heard you can sign whores onto the compound."

"You go ahead. I ain't messin' with that stuff."

"What do you think the third assignment will be?"

"I dunno. It's gotta be a get-over assignment like the other two."

"Yeh, I guess that's about right."

Personally, the idea of mounting guard details for two weeks was dreadful, but as their leader, I felt I had to want what they wanted, even if it didn't appeal to me.

On 23 April, we received a re-supply. Included was the wonderful sight of an overstuffed orange bag containing mail. Santa with his bag of gifts couldn't have been more welcome. We hadn't received any mail in about a month. Sergeant Rodriguez broke it all down by squad and the squad leaders distributed it to the soldiers. Everyone got mail, which provided a significant boost to our morale. I received letters from my parents and from my wife, who was now six months pregnant.

"Wow, LT, we got beaucoup mail, huh?"

"Yeh, that's all right, isn't it?"

"I don't know if I can fit it all in my ditty box."

Every soldier carried a water-tight ammunition can, referred to as a 'ditty box', in which they kept their stationery and letters from home, which were typically read and re-read.

I received instructions from Captain Thomas to move to a point in the valley to our south, from which we would be extracted on 26 April, three days hence. Looking at my map, I found that there was a stream which would take us down out of the mountains to a spot where we could establish a PZ.

The stream was typical of most mountain streams in the area—very rocky, with the rocks worn smooth by the fast-moving, cool, clear water. In many spots there were boulders as big as a car.

Following the stream, we came to a spot where it spilled over a ledge, creating a beautiful waterfall about seven feet in height. It was mid-day and we hadn't bathed in two weeks, so it seemed a perfect time to stop for lunch and a bath. Halting the platoon, I called for Sergeant Rodriguez to come forward.

"Let's form a perimeter here. Make sure to cover the trail we came in on. Fifty per-cent alert. Set up a rotation so that everyone gets a bath."

As in all things, the leader takes care of himself only after his men have been taken care of. While waiting my turn, I watched with amusement as my men enjoyed this unexpected bounty provided by Mother Nature. What was most amusing was watching soldiers on guard while awaiting their turn, lying prone with rifles at the ready, buck naked except for helmet, boots, and weapon.

When my turn came, I stripped down and gingerly stepped into the stream. The water was cold and refreshing. Reaching the waterfall, I stepped down into a rock basin, the floor of which was worn smooth. Standing thigh-deep in the pool, I let the water pound down on my head and shoulders, washing away dirt, grime, and care. Mere words cannot convey what a wonderful, refreshing experience this was. I've had many a bath and shower in my seventy years, but none even remotely compares to this. Sadly though, all good things must come to an end.

"Awright, saddle up."

On 25 April, we reached the valley floor and set up a perimeter in preparation for extraction the following day. The excitement over our upcoming two weeks of easy duty was uppermost in everyone's mind. Suspense was created by not knowing exactly what that easy duty would be. Would it be 95th Evac? Would it be 24 Corps Headquarters? Or would it be the wild card, whatever that might be. With all the impatience of kids on Christmas Eve, we waited to see what the next day would bring.

SP4 Jesse "Blooper" Brown

PFC James "OB" O'Brien

TWILIGHT IN THE 'NAM

The Wild Bunch L-R Unk, PFC O'Brien, PVT Warner, SP4 Elzey, PFC Jackson

37. TWO WEEKS EASY DUTY?

As scheduled, we were extracted from the field on 26 April. As soon as we touched down on the chopper pad at Camp Crescenz, I jumped off and made a beeline for Tom Mylan, our new XO (thankfully him and not me) who was standing at the edge of the pad. He seemed as happy to see me as I was to see him and get the news of our new assignment.

"What's our assignment, Tom?"

"Buzz, I got a really good deal for you. They're building a new firebase and you're going to pull security for it. They've been having lots of contact and…."

Throwing my helmet on the ground, I exploded. "GOD DAMN YOU, YOU SONOFABITCH, YOU FUCKER!"

I continued to berate him about the unfairness of it and how he had no business deciding what assignment I would like.

Poor Tom. He was gob-smacked. He thought he was doing me a great favor (and, in retrospect, he was) but it didn't seem so at the moment. He didn't know how to respond to my outburst and just stood there, looking bewildered and crestfallen. For my part, I was completely blindsided because this was so opposite to what I expected. I was convinced that we were in for two weeks of cushy, safe duty in the rear and instead we were being immediately returned to combat duty.

Two deuce and a half's (two-and-a-half-ton truck) were waiting to whisk the platoon off to the firebase without even a chance to shower or get clean clothes. I was unable to accompany them as I had to go to battalion headquarters for a briefing on this new mission. At the battalion operations section, I was given my new orders.

"Brigade is establishing a new firebase off the east end of the Ridgeline. The base will consist of one battery of 105's (105 mm howitzer), currently Battery B, 3d of the 82d Artillery. There's been quite a bit of contact there. Last week, two NVA were killed with six AK-47's and six rucks recovered. Your mission is to build the defenses for the base and to provide all-round security. You will be attached to the battery. I think the battery commander's name is Captain Pettigrew (pseudonym). Do you have any questions?"

"No, sir."

"OK. I'll round up a quarter-ton to run you up there."

"Right, sir."

Part of the code of the military leader is a maxim that orders are to be accepted unquestioningly and transmitted to subordinates as if they were one's own. I was soon to discover the wisdom of this mandate.

The ride to the firebase followed the familiar route up to the Ridgeline, which we had occupied in February. Once on the ridge, instead of turning left in the direction of my old positions, we turned right and headed east. At the very end, where the ridge began to descend, there was a knoll that jutted out to the north. The top of the knoll had been worked over with earth moving equipment, clearing it of vegetation and digging positions for five 105 mm howitzers as well as the Fire Direction Center (FDC) from which the battery commander exercised command and control. The base occupied an area approximately fifty meters wide and seventy-five meters long, sloping gently away to the north. There were three guard towers, one at the north end and two more at the southeast and southwest corners. A single coil of concertina barbed wire surrounded it and a gate had been erected where the ridge road entered the firebase.

As we passed through the gate, I noted that my platoon was huddled at the edge of the base about twenty yards away. My first priority was to report to the battery commander so I passed by the platoon without speaking to them. The jeep dropped me off near the center of the base at the entrance to a subterranean bunker with a howitzer on top. Entering the bunker, I paused for a moment to let my eyes adjust to the gloom. To the Non-commissioned officer present I said, "I'm Lieutenant Sherwood, Charlie, One-Four-Six, I'm supposed to report to Captain Pettigrew."

"Right, sir, stand by one."

The NCO disappeared into another room in the bunker and reappeared with a young captain with an open, friendly face.

"Lieutenant Sherwood reporting for duty, sir."

"How are you doing, Lieutenant Sherwood? Glad to have you."

"What's the first priority, sir?"

"Well, the first thing you better do is see to your platoon. I think you may have a mutiny on your hands."

Hopefully the dim light hid the flush that must have come over me.

"What do you mean, sir?"

"Ever since they got here they've been sitting over on the perimeter, refusing to move or do anything."

"I'll take care of it, sir."

"Good, then we'll talk about what needs to be done."

"Roger that, sir."

Saluting, I did an about face and re-emerged into the bright sun and scorching heat, regretting my childish outburst at Tom Mylan, observed by all my men and giving them license to balk at this new assignment. I was angry with myself but not at them. If this had happened early in my tour, I would have been terrified that I did indeed have a mutiny on my hands and that they would refuse to obey me, but we had been together long enough to be like family and I wasn't overly concerned. But still, what if…?

I strode across the base to where they were sitting. "C'mon you guys, let's get our asses in gear. We've got work to do."

I halfway expected serious griping and reluctance, but they just got up and secured their gear. Not with any great enthusiasm, mind you, but without any hesitation. And so began the most difficult, and in many ways most rewarding, two weeks of "easy duty" that I ever pulled.

38. FIRE BASE

Having gotten the platoon on their feet and moving, it was important now to get them doing something. I hadn't received any specific orders from Captain Pettigrew but I knew our basic mission was to secure the base, so establishing a perimeter was the obvious place to start. I gathered the squad leaders and platoon sergeant and assigned squad sectors, with additional instructions to cover all dead space with grenade launchers and to place the machine guns where they could cover the most likely avenues of approach. Adjustments could be made later but the most important priorities were to establish a defense and get the men productively engaged. I walked the perimeter until I was satisfied that everyone was working with a will, then I went back to the FDC to get specific instructions from Captain Pettigrew.

Before sitting down, Captain Pettigrew introduced me to his XO, Lieutenant Faulkner. Of lean physique and medium height with sandy, close-cropped hair, he exuded a no-nonsense professionalism coupled with an outgoing, friendly personality. I noted with satisfaction that he was a Ranger, which meant that he had completed the Army's toughest training and would be familiar with our mission as infantry. Captain Pettigrew opened the conversation.

"As you can see, the battery is in place but almost nothing else has been done and all of that falls to you. Unfortunately, almost everything that needs to be done is top priority: security, fighting positions, wire obstacles."

"Yes, sir, plus we'll need to send out patrols, ambushes, and LP/OP's (Listening Posts/Observation Posts)."

"Good, you've got the idea. I have all the engineer supplies you need: concertina, barbed wire, stakes, trip flares, sand bags, etcetera. If you need anything else, just let me know. There are also lots of culvert halves for covering your fighting and sleeping positions. Lieutenant Faulkner has sensors and other toys that will be helpful and he will be glad to be of assistance. Figure out your plan and priority of work and brief it to me and we'll go from there. You can bunk here in the FDC, there' a cot just behind the wall there for you. Any questions?"

"What are the chow arrangements, sir?"

"You'll be eating with us. We get three hots a day delivered from the rear."

"Roger that, sir. I'll get back to you ASAP with my plan."

After Captain Pettigrew left I spoke briefly with Lieutenant Faulkner and we agreed to get together the following day to collaborate in setting up the defense. I instinctively knew that I was going to like working for Captain Pettigrew. He treated me as a fellow professional and saw no need to resort to a bullying or an authoritarian attitude in order to establish his authority. I secured my ruck and deposited it on the canvas army cot behind the plywood wall, a luxurious accommodation compared to the jungle.

Back outside, I again walked the perimeter to satisfy myself of the positioning of the men and then briefed the platoon sergeant.

"Have them start digging two-man fighting positions. One man on guard, the other digging. Make them chest deep and no wider than these culvert halves. Place a culvert half on top for overhead cover. Use the excavated dirt to fill sand bags for a parapet and overhead cover. Also, put another culvert half at the rear of the fighting position to shield them from direct fire by the 105's. Have each squad leader prepare a range card. Make sure all the fires are interlocking, especially between squads."

With the immediate defense of the base being seen to, I turned my attention to the outer defenses. I analyzed the terrain around the base and determined what I thought were the most likely and least likely avenues of approach for an attacker. I decided to try to channel an attacking force into ground of our choosing where we would be able to mass our fires. To do this I would construct the wire obstacles so that they were most dense where I didn't want the enemy to come and less dense where I did want them to come, angling it so that an attacking force would follow the path of least resistance and be nudged into our kill zone. But before that we had to work from the ground up.

First of all, the base would be ringed with claymore mines with overlapping blast zones. Then, tangle-foot would be placed at ankle level all the way around the base to a depth of about 30—40 meters. Tangle-foot consisted of barbed wire strung tightly in an irregular checkerboard pattern at ankle level, staked down with heavy iron stakes. Within the tangle-foot, trip flares would be liberally placed, rigged so that they would go off if tripped or if the trip-wire was cut by a sapper. Once that level was complete, it would all be topped off with three layers of concertina razor wire, also staked down.

After making a sketch of the proposed plan, I briefed it to Captain Pettigrew.

"This looks good Lieutenant Sherwood. How soon can you get it done?"

"Well, sir, we're pretty close to finishing our individual fighting positions. It's getting pretty late in the day to start on the wire obstacles, but we can get started first thing tomorrow."

"OK, sounds good. I'd like for you and Lieutenant Faulkner to work together on this. He has some pretty neat gadgets to add into the mix."

"Roger that, sir."

After being dismissed by Captain Petigrew, I got together with Lieutenant Faulkner.

"Captain Pettigrew said you had some gadgets to incorporate into the perimeter defense."

"Yeh. Hang on a sec."

He disappeared and returned shortly with something that looked like a weed, about two feet tall.

"Check this out."

"What the hell is it?"

"It's a sensor. It picks up ground vibrations, like if someone is walking nearby. They are battery operated and we can monitor them from the FDC. If anyone is in the wire we will know it and bring pee on them."

"Wow, that's really cool. We're going to start tomorrow on the wire. I'm going to take out a patrol in the morning. After that, let's you and I walk the perimeter and decide where to place these babies."

The last thing to be done was to establish security for the night. Getting with the platoon sergeant, I gave the following instructions: "Set up a guard at each tower starting at twenty-one hundred hours. Two-hour shifts, two men per shift. SITREP every hour on the hour. Stand-to at Zero five hundred."

"Roger that, LT."

Before long the dinner meal was delivered in thermal mermite cans. This new assignment was going to be really tough but it did have perks. Getting hot chow every day was certainly one of them.

Later, at twenty-one hundred hours, I walked the perimeter to make sure that the guard was in place and that all was well. Satisfied, I returned to the FDC to get some sleep. To the NCO on duty I said, "I'm going to log some rack time. Would you wake me at Zero-one-hundred? I need to check the guard."

"Will do, sir."

Lying down on the cot, I quickly fell asleep.

BOOM, BOOM, BOOM! I suddenly awoke to loud explosions. Looking around, the dim light of the FDC was further obscured by dust pouring from the ceiling. It reminded me of those World War I movies with soldiers in their dugouts under artillery bombardment. I quickly realized that the battery was firing a mission and that the howitzer above the FDC was creating the immediate commotion. I was soon to learn that fire missions at any and all hours of the night were a regular occurrence and that we would just have to get used to it.

Walking the perimeter at 0100, I found all guards awake and alert. In spite of our rocky start earlier in the day, we seemed to have fallen into the rhythm of our new assignment.

39. MYSTERY PAJAMAS

Since all our tasks were top priority, all would have to be done simultaneously. Accordingly, I set a detail to building the wire obstacles under the supervision of the platoon sergeant, I would take a patrol out to recon the area around the firebase, and the remainder of the platoon would continue to improve individual fighting positions. After those tasks were accomplished, we could then turn our attention to digging sleeping positions.

The morning weather was scorching hot under clear skies and those on wire and digging details were stripped to the waist. For those of us going on patrol, we were in full field gear but, thankfully, without our heavy rucksacks.

I briefed the patrol.

"We're going to go down the hill and scope things out for any signs of dink activity. We'll also be looking for places to set up ambushes. There's been contact here within the last week so we need to be on our toes. No talking. I can see from the map that there is a trail at the bottom of the hill. To the west, it leads out into the foothills and rice paddies between us and the mountains. To the east, it leads to a small village located about four hundred meters northeast of the firebase. The village is defended by Ruff Puffs (RF/PF Regional Force/Popular Force) and is off-limits to us. Any questions?"

There being none, I continued. "Warner's on point, Bugge, you walk slack. Make sure you have a buckshot round in the chamber. I'll be right behind Bugge. Everyone lock and load. All weapons on safe except the point man. Let's go."

After winding our way down the hill, we intersected with the trail, which was well worn and obviously used frequently. To the north there was a stream that paralleled the trail. Unlike the barren hilltop, the bottom of the hill was shaded with plenty of trees and other vegetation. It was very hot and humid. Silently, we crept along the trail, moving east in the direction of the village. We hadn't gone very far before Warner put his left fist up, signaling the rest of us to halt. I went forward.

"What's up?"

"Check it out, LT."

There, on the right side of the trail, was a set of Vietnamese pajamas completely drenched in blood. The blood was wet, red and fresh, seemingly left there within the last few minutes. This was a major clue, but to what? Was the victim VC or a villager? What caused the wound? Were they heading toward or away from the village? We had heard no small arms fire. We periodically fired grenades into the defensive zone around the firebase, but, other than that, we had not fired our weapons. The trail was hard-packed and dry, so it was impossible to detect fresh footprints and a direction of movement. Looking around the immediate area, we could find no further clues such as a hastily dug grave or any other items of clothing.

"Keep moving down the trail. Take it slow and be extra alert. Keep an eye out for blood trails."

"Roger."

After reporting in over the radio, we continued to move cautiously down the trail in the direction of the village. There was no blood trail and no further signs to help unravel the mystery. Stopping short of the village, we backtracked and returned to the firebase. By now it was early afternoon and blazing hot. I tracked Lieutenant Falconer down in the FDC.

"You wanna go put those sensors in now?"

"Yeh, let's do it."

Accompanied by a couple of his soldiers, we made our way around the outer edge of the wire obstacles, emplacing the sensors along routes where an attacker would have to go. We took care to put them in the midst of other vegetation so that they would look just like any other weed. Lieutenant Falconer made careful note of where they were located so that he could direct fire on them if they were activated.

"These things are really cool. I can't wait to see how they work out."

For the rest of the afternoon I checked on the progress of the details building wire obstacles and digging fighting positions, helping out here and there with stringing wire and filling sandbags. Morale was good.

In the evening a jeep came up from the rear bringing hot chow and, another unexpected bonus unheard of to us infantrymen, cold sodas. The work was very hard but there were benefits.

After dinner came yet another surprise—a movie! I had some reservations about showing movies in a situation where we could be attacked at any time, but I wasn't in charge. I made sure that the guard was set, assuring those on guard that they would get their chance to see movies on subsequent evenings.

The evening's cinematic offering was 'Love Story', an immensely popular movie that I had seen with my wife on our honeymoon in exchange for her going to see 'Patton'. It starred Ryan O'Neil and Katherine Ross.

On a previous screening, the film had broken and the broken segment was spliced back onto the wrong reel, the result being that we watched the end of the movie first. As the film began, we watched the final scene of Katherine Ross dying. Having been denied the tear-jerking prelude to her death, there was no sympathy.

"Awww, she's croaking off already?"

"What a waste of a good-looking babe."

"What a rip-off."

"Being in love is never having to say you're sorry, what a crock."

"Sorreee."

Having already seen the end, the rest of the movie was anti-climactic, producing nothing but guffaws and ribald comments. But it was entertaining, if not in the way it was intended.

We saw lots of movies while on the firebase, most of which are long lost to memory. One that I remember, because it was so unlike anything I would have ever voluntarily gone to see, was 'Two Lane Blacktop', a roadie movie about two guys driving across the country on Route 66, drag racing along the way in order to make money.

The hands down favorite was 'The Legend of Billy Jack', a story about a Native American Vietnam veteran who goes back to his hometown in the Southwest and takes on the corrupt and racist power structure there.

After 'Love Story' was over I made the rounds to check the guard. Turning in, I again asked the Duty NCO to wake me at 0100 so that I could check the perimeter.

"LT, it's zero one."

Sleeping fully clothed and never more than half asleep, I immediately sat up, swinging my legs off the side of the cot.

"Thanks," I said as bent over to put my boots on.

Stepping out of the FDC into the hot night air, I waited for my eyes to adjust to the dark and then made my way to the tower at the north end of the base.

"Pssst. It's me, LT," I whispered as I climbed up to the tower, "How's it goin'?"

Brown and Tetrault were on guard.

"Hey, LT. Everything's quiet."

"Good. No movement?"

"No, nothing."

"Go ahead and chuck a hand grenade down range every now and then. Just make sure you throw it past the wire so that you don't set off any trip flares and mess up the wire."

"Roger that."

I thought this would give the VC second thoughts about snooping around the perimeter and would be a good way to keep the guards awake and alert.

I visited each guard position and hung around for a while before making my way back to the FDC for a few hours' sleep, interrupted every now and then by the cannons firing. All in all, things weren't so bad.

40. SETTLING IN

Had I been more mature or reflective, I might have pondered my enthusiasm for this new assignment, especially in light of my anger at having received it in the first place. Perhaps I would have thought about the fact that we had a mission that produced tangible results. We were building this firebase from scratch. It was rising up from the ground by the work of our hands. It was ours. Even though our mission was one of defense, it wasn't static and boring as the Ridgeline had been. Ours was an active defense, with patrolling and ambushes. The recent contacts held out the hope that we could finally get the drop on Charlie. Of course, receiving hot rations and entertainment wasn't bad either. Finally, Captain Pettigrew, my temporary commander, treated us collegially and with respect.

Although the fighting positions and wire obstacles required constant maintenance and improvement, they were, for all intents and purposes, complete. Our attention now turned to the digging of sleeping positions. Behind each fighting position we had to dig a two-man sleeping position. Each position had to be about seven feet long, four feet wide and about five feet deep. The top of the position had to be flush with the ground so that, if necessary, the cannon could fire point blank. Lots of digging. Once the holes were dug, dirt-filled wooden ammunition cases were used to line the walls. As with the fighting positions, culvert halves were used to create the roof, and this covered over with several layers of sandbags. While these positions were good for stowing gear out of the heat and rain, the troops found them uninhabitable, preferring instead to sleep on the ground on top of them.

"O'Brien. How come you guys aren't sleeping in the new positions?"

"It's too hot, LT. It feels like you're roasting in there."

"Well, at least it's there if you need it."

While I helped out here and there with digging and filling sandbags, my main focus was on the overall defense, spending most of my time leading patrols, laying ambushes, and checking the perimeter with Lieutenant Falconer, looking for signs of tampering.

In the movies, ambushes look very exciting, with the ambushing element getting into position, waiting a few minutes for the unsuspecting enemy to come along and then mowing them all down in a hail of gunfire. The reality is that most ambushes are a dry hole, producing nothing. And that was after hours of lying silently with the elements, trying hard to stay awake and focused.

The trail at the bottom of the hill was the main object of our patrolling and ambushes, since it was the conduit between the village and the mountains. One day I set an ambush on the trail, initially with the highest hopes of bagging Charlie. Nestling into the undergrowth with our weapons aimed at the trail, we waited. After a couple of hours of laying in the bushes in the heat of the day, I struggled to stay awake.

God, I can hardly keep my eyes open. Stay awake. God it's hot. I'm soaked in sweat. I'd like to get a drink from my canteen but it might make too much noise. I hope everyone is awake. I need to take a leak, but I can't. I wish Charlie would come diddlybopping down the trail. What's that tickling my arm? HOLY SHIT!

Glancing over at my right arm to see what was tickling it, I discovered several scorpions crawling around on my forearm. My first instinct was to recoil and try to brush them off but I caught myself, realizing that their reaction to any such disturbance would be to sting me. They were just checking things out and had no reason to sting me as long as I didn't provoke them. By staying as still as possible, they eventually left, much to my relief. Charlie didn't come, so I terminated the ambush and we headed back to the firebase.

One night for the evening's entertainment, the Battery held an all-comers boxing match, what is known in Army parlance as a "Smoker". Each pair of pugilists would square off for three three-minute rounds wearing twelve-ounce boxing gloves.

"Look at those two. They can't fight for shit."

"Yeh, it looks like two windmills."

"Hey, open your eyes."

"You gonna go out there?"

"Fuck no. I'm a lover, not a fighter."

"Right."

By round three, neither contestant could keep their arms up and they staggered to a finish, so to speak.

Inevitably, there would be a mismatch between someone who thought they knew how to fight, and someone who knew how to fight.

"Now these two look like they know what they are doing."

"Yeh, but that big guy has got a lot of reach on the Chicano dude."

"Let's just see."

As soon as they squared off, the Mexican feinted with his left, causing an opening for a powerful right hook, followed by a flurry of well-aimed punches to the face and head. The big man went down in a heap in less than thirty seconds.

"What's that you were saying about the big man?"

"Aw fuck you."

In addition to patrols and ambushes and sensors, I also employed Listening Posts (LP) at night as an early warning device. The main LP was situated about three hundred meters to the northeast between the firebase and the village. Other LP's were placed as the situation seemed to dictate. The LP consisted of two soldiers with rifles and a radio. Their job was to warn us in advance if anyone was heading our way, a very dangerous assignment. Not having enough firepower to defend themselves, they had to remain completely silent. The radio watch on the firebase would contact them hourly throughout the night.

"LP, this is Charlie one-zero Romeo. If you have negative SITREP, break squelch twice."

KSHHHHT, KSHHHHT.

Depressing the transmit button on the radio handset made a 'KSHHHHT' sound. If all was quiet, the soldier on the LP radio would depress the transmit button twice, indicating that everything was alright. The only time the LP was allowed to actually talk on the radio was in the event of an emergency.

Our life as infantrymen was punctuated at all hours of the day and night by the artillery firing missions.

"FIRE MISSION!" bellowed the Chief of Smoke (Fire Direction Sergeant).

"FIRE MISSION!" echoed the crews on the guns. Each fire command was echoed back to the Chief of Smoke by the crew to ensure that they had heard the command correctly.

"TROOPS IN THE OPEN."

"SIX ROUNDS, CHARGE THREE."

"DEFLECTION 3111, QUADRANT 400."
"FIRE!"
BOOM, BOOM, BOOM went all five cannon until the mission was complete.

Before coming in contact with the "red legs", the only soldiers we had any respect for were other infantrymen. However, the willingness of the red legs to spring into action at all hours to fire in support of the infantry caused us to add them to the circle of soldiers worthy of respect. A sign of this respect was manifested by the grunts mimicking the artillery firing commands when firing grenades around the perimeter. I had instructed my men to periodically fire or throw grenades at the outer edge of the wire perimeter in order to discourage snooping. Now, you would hear the squad leader barking the fire commands in imitation of the Chief of Smoke, with the grenadier echoing it back. Instead of a thunderous boom, now you would hear the hollow, metallic "Bloop" as the grenade left the tube.

And so we settled into a routine that was exhausting but not without gratification. Patrolling, ambushing, and building by day, ambushing, listening, and inspecting the guard by night.

Firebase

Firebase, North Tower

Firebase, Fire Direction Center

Firebase, Cannon crew

Firebase, Sleeping positions under construction

41. ARTILLERY RAID

With our grueling schedule, the days sped by and, before I knew it, we'd been on the firebase for over two weeks. Our work became much harder when, after a few days, I received a message from Captain Thomas requiring me to provide three men to fill the guard quotas at 24 Corps and 95th Evac. At first I welcomed this as an opportunity for at least a few of my soldiers to enjoy the easy duty they had so looked forward to. But, as the days progressed, I continued to get requirements to send more men, until our strength was down to twenty with no relief in our duties. Captain Thomas had never bothered to come and check on us nor would he ever.

On the 14th of May, Captain Pettigrew announced that the battery was going on a raid the following day.

"What's an artillery raid?"

"It's when they fly the battery to the field to fire Direct Support for a specific infantry mission, then we fly back."

"So what mission are you firing in support of?"

"A company of Two-one is sweeping down a mountain where there is a suspected NVA unit. Ruff Puffs will set up a blocking position at the base of the mountain. We'll set up in the valley and provide fire support."

"Wow. That sounds pretty cool. Mind if I tag along? I can help with security."

"Sure, come on along."

The following morning dawned clear and hot. The gun crews were busy rigging their guns to be slung beneath the huge Chinook helicopters that would ferry us to the field location.

As the first Chinook arrived, creating an enormous dust storm, I clambered aboard with the gun crews. The rear ramp closed and we lifted off, but only to hover over one of the guns as it was hooked up to the big cargo hook on the Chinook's belly. Once that was completed we took off in earnest, heading southwest toward our destination in an area known as the Arizona Territory, so named because of the wild and wooly fighting that had taken place there throughout the war.

After flying for about twenty minutes we began our descent. Hovering again, the gun was eased to the ground and the sling strap was released. Then we slipped sideways a few yards and landed, the rear cargo ramp was opened and we piled off. The gun crews immediately went to work manhandling their gun into position and prepared for firing. In a steady progression over the next few minutes, the same process was repeated as the remaining four guns arrived.

Once the battery was in position, they began firing registration rounds to ensure that their fire, when called for, would be synchronized and on target. As I watched, I was pleased to note that the red legs had brought coolers full of ice-cold coke. It was super hot. I walked over to one of the coolers and asked of the soldier standing there, "Mind if I have a coke?"

"Help yourself, LT. There's plenty."

"Thanks," I said as I reached deep into the cooler and grabbed a can. Opening it, I savored the icy burn as the brown liquid went down my throat.

We were in a broad, flat valley that was hemmed in to the north, south, and west by mountains. The operation we were supporting was at the northwest corner of the valley, about five hundred meters away.

As I surveyed my surroundings, the Ruff Puff unit was moving past, heading to their blocking position at the base of the mountain. They were a motley looking group, dressed in typical peasant pajamas, with rubber flip flops for footgear. Some had fashioned backpacks out of metal wastebaskets, using twine for shoulder straps. But they were armed and hopefully were ready to do the business.

It was impossible to see what was going on with the ground operation, as the infantry were invisible under the triple canopy forest that covered the mountain. As the day wore on, their progress could be gauged only by the shifting of the artillery fire working its way down the mountain.

The guns were all registered and ready to go but it took several minutes for the Ruff Puffs to make their way down the valley and get into position. The day was getting hotter and hotter as the sun rose in the sky. The gun crews were stripped to the waist but I remained in full uniform. I had another coke.

Once the Ruff Puffs were in position, the pace of the action started to pick up. That is not to say fast, because moving through mountainous, jungle covered terrain is never fast. But the big guns were steadily firing. Since they had direct line of sight to the target area and we were relatively close, they were firing in the direct fire mode rather than indirectly, as was usual for artillery. At first they were firing high explosive rounds, which we could identify by the pall of dirty brown smoke that rose from the jungle. As the tempo of the operation intensified, the guns started firing a combination of high explosive (HE) and white phosphorous (WP, Willie Pete, or Wilson Pickett on the deck, in the parlance of the GI's), which quickly shrouded the entire mountainside in thick white smoke. I didn't know why they were firing the WP and I couldn't imagine what it was like for the grunts on the ground with dense clouds of smoke further obscuring their vision.

From the site of the operation, a Huey helicopter flew low across the valley floor in our direction. As it reached our location, it came to a hover and dropped off the body of a Ruff Puff who had been killed. This should have produced some emotional response in me, but, amidst all the noise, firing and activity, this was just one more event impartially observed.

'BANG'

A loud explosion about twenty meters north of our position. Thoughts raced through my head, *What the fuck was that? The dinks must be attacking us. Get ready.*

Looking around, I noticed an unmanned M-60 machine gun in the direction of the explosion. I ran over to the gun, lay down, pulled the charging handle to the rear, lifted the feed tray, inserted a belt of ammunition and prepared to fire. Everyone else who wasn't actively engaged firing the cannons was also grabbing their weapons and preparing for an assault. We waited. Nothing happened. Gradually everyone resumed their former activities. I had another coke.

As the day tipped into the afternoon, the sky began to get cloudy, progressing from a few puffy white cumulus clouds to a dense, gray blanket, which covered everything. The heat transformed from blast furnace hot to suffocating hot. I had another coke.

The guns continued to bang away all afternoon, the impact of the rounds creeping downhill ahead of the advancing infantry. We could occasionally hear the crackle of small arms fire in between the artillery explosions.

The day eventually came to an end without further incident or revelation, other than a vague report that the infantry had found an NVA hospital. Such is often the way in war. One plays a small role in a much larger scheme, quite often never finding out what the big picture was and how their role fit into it. I think this was particularly true in Vietnam, where it was about destroying the enemy rather than seizing and holding terrain.

The guns went out of action. The gun crews swabbed the tubes and prepared them to be slung out. The Chinooks arrived and we repeated the morning process of loading the crews and then slinging the guns underneath. It was getting dark as we began the flight back to the firebase. Under stormy skies, the wind was blowing, causing the cannon slung beneath us to oscillate, which in turn caused the whole helicopter to rock unnervingly back and forth like a boat in choppy seas.

Arriving back at the firebase around eight p.m., it had started to rain quite heavily. I was dead on my feet but, it being my mother's birthday, I felt compelled to write her a letter. The past two weeks had been so busy that I hadn't written any letters but I couldn't let the day pass without writing her.

After penning a short letter, I lay back on my bunk. All of a sudden, I felt an urgent, ominous grumbling sensation deep in my bowels. Hurriedly I got up and headed for the entryway to the FDC. I took about two steps and my bowels completely gave way. Those cokes I had drunk all day had liquefied my insides. I was mortified. I'd lost control. Men don't lose control. Men don't soil themselves. I slunk out of the FDC unnoticed. By now the rain was a torrential downpour. I went to the makeshift shower that had been rigged up and cleaned myself and my uniform as best I could. Re-dressing in my sodden clothes, I slipped back into the FDC. Fortunately, I had gotten an extra uniform sent out from the base camp so I was able to change into something dry. Before turning in to get some sleep, I poked my head through the curtain and said to the duty NCO, "Sarge, wake me at zero one hundred, will you?"

"Sure thing, LT."

Artillery raid, white phosphorous smoke on the target area

Artillery raid, 105mm howitzer fring

42. INSPECTION AND HO CHI MINH'S BIRTHDAY

On the eighteenth of May, three days after the raid, the firebase was inspected by the brigade commander, Brigadier General (BG) Joseph McDonough. In the Army, being inspected by a senior officer is always a source of anxiety. No matter how well prepared one is, there is always the fear that fault will be found. I was doubly apprehensive due to my last encounter with General McDonough, when he read me my rights. On the other hand, I was proud of the work we had done and eager for it to be seen and acknowledged.

The visit began with a promotion ceremony for a senior NCO, after which the inspection tour began. Captain Pettigrew led it off, showing the CG (Commanding General) the FDC and gun positions, introducing him to members of the battery, and describing the various missions they had fired. At this stage of the inspection, I was more in the role of a bystander, and as I observed BG McDonough, the wheels of my memory began to turn.

Joe McDonough, Joe McDonough. Dad used to always talk about his friend Joe McDonough from the Korean War.

"I wonder what ever happened to Joe McDonough? he would say, "He was a West Pointer and a really good officer but he wasn't sure he was going to stay in the Army. It would be nice to find out what became of him."

As I looked carefully at BG McDonough, I began to recognize the face in the snapshots of my father's from Korea. He was twenty years older and gone bald, but beyond that I could clearly see that this was my father's long lost friend. I couldn't wait to write and tell him of my discovery.

Meanwhile, the tour of the battery came to an end and it was my turn to show the CG the infantry positions and defensive plan.

"Sir, these are our fighting positions. Each two-man position has interlocking fires with adjacent positions, as you can see on my fire plan. My machine-guns are located here and here to cover the most likely avenues of approach, and my grenadiers are located here, here, and here to cover the dead space to their front. The wire obstacles have been constructed so that an attacking force will be channeled into the most likely avenue of approach, which is also our kill zone. "

The general nodded approvingly, so I continued. "Each fighting position has a shield in the rear to protect the soldiers from direct fire from the howitzers. Behind each fighting position is a two-man sleeping position like this one."

Again an approving nod and a question. "This looks good. You've done a lot of work here. How do the troops like these sleeping positions?

"Sir, the troops generally sleep on top. It's too hot inside."

"I see. What else do you have to show me?"

"There's not much more to actually show you, sir, but I would add that we conduct daytime patrols and ambushes and at night we maintain at least one LP/OP. And, we maintain guard positions on the perimeter during the hours of darkness."

"How many men are on each guard post and how long are the shifts?"

"Sir, initially I had two men per post with a two-hour shift. Now that I am down to twenty men, I've reduced it to one man guard posts."

"Bullshit! You always keep at least two men on guard at any one time, no matter how many men you have."

"Yes, sir."

"Why are you down to twenty men?"

"Sir, I've been required to send men to the company to fill out the guard at 24 Corps and 95th Evac."

"Well Lieutenant, it looks like you are doing good work here. Keep it up. And make sure you put two men on guard."

"Yes, sir. Thank you, sir."

All in all, the inspection seemed to have gone well, in spite of the rebuke over the number of men on guard, which came across more as fatherly than harsh.

That night I wrote my parents and told my dad about my discovery. I wanted to extend his best regards to General McDonough but was concerned about causing him embarrassment due to my previous run-in over the donnybrook at the air base. I decided to leave it in his hands. I also expressed my excitement at the prospect of becoming a father in two months and hoping I would be allowed R&R for the occasion. This was not a given. Captain Pettigrew's wife had just given birth and his request for R&R was denied.

The next day, 19 May, was Ho Chi Minh's birthday and word came down through intelligence channels to expect an attack that night. I gathered the platoon sergeant and squad leaders to brief them.

"OK, listen up. Today is Ho Chi Minh's birthday and the G2 is that we may get hit tonight. Double check each fighting position and make sure there are plenty of grenades and ammo. Check the wires on all Claymore's and test the connections. Sergeant Fenstermacher, take a patrol out and check the outer perimeter for any signs of tampering with the wire. Sergeant DiRinaldo, you take a patrol out and check out the trail and all around the base of the hill. Look for any signs of increased activity or preparation for an attack. Check out any new trails off the main one. Look for possible assembly areas or cache sites, anything out of the ordinary."

"Yes, sir."

"Any questions?"

"No, sir."

"OK. Let's get ready for these fuckers. Tonight, we'll be on one hundred percent alert."

The rest of the day was spent in a high-energy flurry of activity. Sergeant Rodriguez and I checked each position for readiness, re-examining sectors of fire and making sure each soldier knew what to do. I double-checked with the FDC to make sure we were in sync with the artillery protective fires. Sergeant Fenstermacher reported back that the outer perimeter was intact and that there were no signs of tampering. Sergeant DiRinaldo reported back that there were no signs of irregular activity anywhere at the base of the hill.

Darkness fell with everyone in a state of eager anticipation. When I say eager, I don't mean to suggest happiness, but neither was there inordinate fear. I suppose the feeling was similar to that of a professional team as they wait for the big game to start. Something important was going to happen and we all knew we needed to be ready for it. I went from position to position, asking if noise or movement had been detected.

But, as the night wore on and minutes turned into hours, adrenaline drained, fatigue set in and the task of staying alert became a force of will. In the end, the night produced nothing more than one exhausted platoon. Chalk up one more dry hole.

43. LIFE ON THE FIREBASE

As directed by the brigade commander, I increased the guard to two soldiers per post. I put myself on the guard roster for the one to three a.m. shift in order to lighten the load a little. One night after checking all the positions, I made my way to my assigned post to pull my shift. My partner that night was Specialist Danny Bishop, a taciturn country boy from Arkansas. We lay on top of the bunker, partially because of the heat but mostly because of the better range of vision to our front. In addition to our weapons, we had several hand grenades laid out in front of us. Bishop had a pocket size transistor radio, which was tuned in to a Vietnamese station with the volume turned down to the level of a low whisper. As we lay there scanning the emptiness to our front and talking quietly, I was entranced by the eerily beautiful sound of the twanging one-string Vietnamese instrument accompanying a woman plaintively singing what must have been a love song. Not wanting to get lulled by the exotic music, I followed my own guidance to those on guard.

"Let's chuck some hand grenades down range."
"Roger that."
"See that draw down there past the wire, where it's real dark?"
"Yeh."
"Good. Let's aim for that."

We each secured a grenade, pulled the pin and let fly.
BOOM. BOOM.

"Good aim, Bishop. Right on target. Hopefully that will make Charlie think twice about trying to sneak up on us, but if nothing else, it'll help keep us awake."

And so we passed our two-hour guard shift, talking quietly and occasionally throwing a grenade out past the wire. As our shift came to an end, Bishop went and got his replacement and then I went and got mine. Back in the FDC, I lay down to get a couple of hours sleep until it was time to be up for stand-to.

Almost miraculously, the men who had been pulled off the firebase to pull guard at 95th Evac and 24 Corps started returning. Sergeant Rodriguez and I marveled at this sudden change in our fortune.

"Whadda you make of the guys coming back all of a sudden?"

"I'll bet the CG ripped someone a new asshole for pulling men from a combat mission to pull some broke-dick assignment in the rear."

"There it is."

"Whatever the reason, I'm glad they're back."

Having everyone back significantly lightened the load.

***

On the night of 22 May, we had an unexpected treat—a floorshow consisting of a magician and three female singers. Instead of the usual Vietnamese or Korean performers, these were from the United States, from West Virginia. They were very friendly and we were all immensely pleased, especially to see American girls. The last time we had seen American girls was when the Donut Dollies visited us on the chopper pad as we were about to go to the field.

***

I may have entertained some doubts about the wisdom of having a show on a firebase, but any such doubts were far overshadowed by the thrill of having this group come out to see us, instead of staying safe and comfortable in the rear, which was what usually happened.

***

One day I was tasked to provide security for a mine sweeping team from the Engineers as they swept a nearby road for mines. The Engineer team consisted of two men with a metal detecting mine sweeper. Although it seemed a very mundane assignment, I readily provided a fire team of four men and an NCO to accompany them. When they returned in the afternoon I asked them if any mines were found. To my surprise, the road was mined. The soldiers, too, seemed pleased that what at first seemed like yet another make-work assignment actually produced results and made their effort worthwhile.

One night a big firefight broke out in the village to the northeast of the firebase. During hours of darkness I always kept a listening post (LP) between us and the village. Their job was to provide early warning of an attack from that direction. Being on LP was a particularly dangerous job because of the possibility of becoming stranded. This night, that seemed to be what might happen as the firefight spilled out of the village and moved in the direction of the LP. Olsen was the senior man on this two-man post. He was an old-timer, dour and steady, but this was a situation to try the best of men.

After putting the platoon on one hundred per cent alert, I contacted Olsen on the radio.

"Lima Papa, this is one zero, over."

"This is Lima Papa, over," came the whispered reply.

"Give me a SITREP, over."

"There's a firefight going on and it's heading in our direction. We need to get out of here, over."

"Hold what you've got. We're on alert and we'll make sure you're OK, Over."

"Roger."

I wasn't sure how I was going to make sure they were OK, but for now I didn't want to risk their safety by having them break cover. The rapid, high pitched fire of the M-16's, answered by the slower, louder 'Klack, Klack, Klack' of the AK-47's grew louder as it advanced in our direction. I surmised that the VC were retreating from the village, pursued by the Ruff Puff's. We couldn't open fire because of our inability to distinguish friend from foe in the darkness. Olsen called in on the radio. His whisper now sounded a little frantic.

"We need to get out of here, over."

There was no tactical reason to keep them in place since they had already fulfilled their purpose. My concern now was for their safety. In their present position, they were hidden and might be missed altogether even if the VC continued moving in their direction. If they broke cover to return to the firebase, they might be detected by the VC and shot or they might simply get caught in the cross-fire. I was madly trying to make the right decision, wanting to support Olsen's wishes but feeling instinctively that staying put was the better course.

"Steady. Just hold your position. Stay hidden. I've got an element ready to come get you if necessary. Hold what you've got."

I knew that wasn't what Olsen wanted to hear but he was a good soldier and he obeyed.

Just when it looked like it was reaching the crisis point, the running gun battle petered out and quiet returned. Now I needed the LP to resume their role of listening, knowing that the VC were in the area. I continued to communicate with Olsen in whispers over the radio.

Tactically, this wasn't a good idea, but I felt that he needed a reassuring voice to calm his nerves. Olsen and his companion definitely earned their pay that night.

***

One day Lieutenant Falconer approached me. "One of the sensors has gone dead. Let's go take a look."

"Sure thing. Let me grab a couple of grunts to go with us."

We exited the firebase through the zig-zag lane in the wire and made our way to the bottom of the hill where Lieutenant Falconer had placed the sensor that had gone dead. The sensor looked like a tall weed and, amongst the other weeds, was hardly distinguishable from them. Outwardly, everything seemed in place. Lieutenant Falconer squatted down and opened the battery compartment. The battery compartment was empty. The batteries had been removed. Inside the battery well was a small, rolled up piece of paper. Taking it out and unrolling it we found printed in crude lettering, 'Fuck you GI'.

"Well I'll be a son of a bitch."

***

In a 23 May letter to my parents, I wrote, "We go back to the bush day after tomorrow and for the first time, I'm not real enthusiastic about going. I'm tired, very tired, and the heat is unbearable. We will have to carry a minimum of 10 quarts of water this time as some of the streams will be drying up... But we are grunts and our place is in the bush. I'm sure I will get used to it again soon enough."

44. VISITORS

In addition to the visits by General McDonough and the USO troupe, we had other visitors to the firebase, mostly toward the end of our time there.

One Sunday, a Catholic chaplain came to say mass. He was relatively young, earnest, and soldierly looking.

Out in the open, on a table made of stacked ammo boxes and a plank, he set up an altar from the items in his portable mass kit. The tablecloth, the cross, his prayer book, and finally the items necessary for communion: the chalice, the plate, the wine, and the bread. Those of us so inclined to attend, about twenty, lined up in imaginary pews, our helmets for seats. The chaplain put on his camouflage stole, invited us to stand, and began with the familiar refrain, "Let us begin our celebration in the name of the Father, the Son and the Holy Ghost."

"Amen," we responded after crossing ourselves.

The high points for me were the Agnus Dei, the Lamb of God, which, when chanting it as a boy, never failed to make my ever-present anger melt away, and, of course, the Eucharist, or Communion. Communion was something tangible, something consumed individually and yet shared communally.

The paradox of worshipping a god of peace in the midst of war was lost on us, but not the peaceful interlude in the midst of our world of violence and unpredictability.

This was the only time in my whole tour in Vietnam that a chaplain came to what might be called the field or a forward position. We had a chaplain for our battalion but I never saw him except when we were in the rear. I don't know what he did with his time, but he didn't spend it with the grunts. He was a nice enough guy but that was about as far as it went. In a letter to my parents I wrote, "…the chaplain is OK but is a dud as a man of religion."

"LT, how come our chaplain never comes to the field?"

"Who knows."

"You think he's afraid?"

"I dunno, but you'd think that a chaplain should be the least afraid of dying."

"There it is."

One time when we were in the rear I was in the little hut that passed as the officer's club. The chaplain was sitting at the bar and I overheard him complaining that chaplains were no longer authorized to receive a certain medal (probably the Air Medal). That was enough for me, I wrote him off.

On another occasion, the man who played the character of Wally on the Ozzie and Harriet show came to see us as part of a USO tour. In my daily briefing to my men I announced that he was coming. "Wally, from the Ozzie and Harriet show is coming out to visit us."

"Who?"

"What's Ozzie and Harriet?"

His visit was kind of a lead balloon. While I sincerely appreciated that he would come out to see us, I found his visit pathetic. He was a middle-aged version of the character he played—paunchy and dumpy looking. Since hardly any of the troops knew who he was, there was little or no star appeal. For all that, he was very friendly and seemed to want to bring some cheer to the men. The chemistry just wasn't there.

Toward the end of our stint, Captain Pettigrew let me know that a civilian reporter was coming out to interview my platoon. This was not good news to me. I had a very hostile attitude toward reporters and the last thing I wanted was for one to come sniffing around us. My opinion was that they already had their story, all they needed was some names and faces to go with it. And the story was this: U.S. troops are an unmotivated, undisciplined rabble who use drugs and avoid performing their assigned duties. I gathered the platoon and briefed them on the visit.

"There's a reporter coming up here and he wants to interview you as a group. I won't be there. We live in a free country and you can say whatever you want. All I ask of you is this... Make sure that whatever you say is something you'll be proud to read on the front page of tomorrow's paper. You remember what happened with Two-one up at Phu Bai."

The reporter came and I linked him up with the platoon, then I stayed away as promised. After he left I asked Sergeant Rodriguez how it went. He said that the reporter kept fishing for the responses he wanted but all he got was comments like, "We just want to kill dinks," and other expressions of gung ho bravado. I couldn't have been prouder of the men, not because of the bloodthirsty language, but because of their collective pride and unwillingness to have themselves and their comrades publicly disparaged. To my knowledge, the story never made the papers, which was fine by me.

45. EUREKA

On 25 May, we headed back to the field, specifically, to Charlie Ridge. All of us were a little sad to depart from our Redleg comrades, whom we had grown to respect and, of course, the relative comforts of the firebase. Personally, I would miss working under Captain Pettigrew.

Before leaving, I stripped my rucksack of everything that wasn't absolutely essential in order to make way for the ten quarts of water each of us would carry on account of the intense heat. Included in the discarded items was my precious "ditty box", the airtight ammo can in which I carried my writing paper, letters and other items that would otherwise be destroyed by the weather.

I also quit smoking. Just prior to our departure from the firebase, I received a letter from my dad telling me that he had been diagnosed with emphysema. In a knee jerk reaction, I decided on the spot to quit smoking. I got rid of my cigarettes, Zippo lighter, and all the waterproof plastic boxes in which we carried individual packs of cigarettes. This was a decision I was soon to regret.

For a smoker, a cigarette is the antidote to many ills, to include fatigue, fear, hunger, stress, and physical discomfort. By day one back in the field, I was experiencing all these and was desperate for a smoke. The decision to resume smoking was easy enough, but finding something to smoke wasn't. It didn't seem right to bum off my men, so I relegated myself to smoking ten-year old cigarettes from our C-rations until re-supply, when I could meet my needs from the abundant supply of tobacco products contained in the Sundry Packs that came with each re-supply.

(At the time, I thought it a super patriotic gesture on the part of the tobacco companies to provide us with all these free tobacco products—cigarettes, cigars, chewing tobacco, you name it. Little did I realize that this was a very clever strategy on their part to create a whole generation of nicotine addicts).

Very shortly after our return to the bush, we began to hit pay dirt. We found several cave complexes containing clear evidence of VC/NVA occupation—ammunition, medicine, cook pots, several pairs of Ho Chi Minh sandals, personal equipment, etc. We also found some ammunition caches, which got us written up in the brigade's daily newssheet. At the same time, 3d Platoon of Charlie Company encountered a booby trap consisting of a 105-millimeter artillery round attached to a trip wire, which they blew in place.

Then, on 3 June, we came upon the biggest prize yet. Patrolling out onto a protrusion from the ridge, the point man signaled to halt. I went forward to where he was crouched. "Dink fighting position," he whispered as he gestured to the front.

Sure enough, there was a fighting position and, as we visually scanned the area, we could see that there were more. No one seemed to be there so we cautiously began to patrol the area, each squad fanning out in a designated sector.

"LT, come check this out."

There, in what was turning out to be the center of an NVA position, was a Rube Goldberg looking device which, I quickly surmised, was a pad for launching rockets. On top of a boulder was the tail fin assembly for a 750 lb. bomb, positioned to function as the cradle or rocket launch ramp. Behind it were two stabilizer braces wedged by rocks and tied together with cloth. The surrounding ground was scorched as were the fins of the tail assembly. Looking in the direction that a rocket would seemingly have been pointed, I found branches placed upright as aiming stakes. I took compass readings on them and, plotting them on my map, found that Firebase Linda was right in the cross hairs, so to speak. Meanwhile, the squad leaders had been reporting their tally on fighting positions found, which now totaled 30 - 40. Battalion had long been looking for the sites from which the NVA had been launching their rockets, so this seemed to be a major find. I got Captain Thomas on the radio. "Charlie-six, this is Charlie One-zero, over."

"Charlie-six, go."

"Roger, Six, we have a rocket launch site located at niner-one–niner-six–three-one, over."

"Good copy, over."

After describing the construction of the launcher, I reported the information on the aiming stakes.

"There are two sets of aiming stakes approximately twenty meters from the launcher on an azimuth of 75 degrees. Another set of two stakes are located twenty meters out on an azimuth of 34 degrees, how copy, over."

"Good copy, over."

"Roger, will proceed to destroy the launcher and search the area for more intel, over."

"Roger."

"This is Charlie one zero, out."

A little while later, after reporting this find up to battalion, Captain Thomas let me know that battalion was quite pleased with this discovery.

We continued to scour the area but didn't turn up much else. Moving on to the south, we set up next to a valley in a beautiful spot that I described thus in a letter to my parents, written on pages torn from my leaders notebook:

Right now we are set up in a beautiful spot—undoubtedly the nicest since I have been over here. It is perfectly flat with short grass—just like an unkempt lawn. Bamboo thickets surround the edge, offering shade and concealment. It is also surrounded by a dike which offers cover in case we get hit. But the real gem is the stream we are set up next to—a big, roaring mountain stream which provides water for drinking and a big deep pool for bathing…

It started to rain so I made a hooch out of my poncho, tying one end to a clump of bamboo and the other to my machete, which I had stuck in the ground. As I was laying under this cover waiting for my squad leaders to arrive to brief them on the following day's activities, I noticed a bright green snake slithering through the bamboo to which my hooch was tied. It was a Bamboo Viper, a poisonous snake that was very much dreaded, said to be a 'two stepper', e.g. if bitten by one, two steps and you were dead. (I later learned this wasn't the case, but we certainly believed it). I was crawling out of my hooch just as the squad leaders arrived.

"There's a fuckin' Bamboo Viper in there."

With our attention diverted from the following day's activities, we crowded around the base of the bamboo, looking for the snake. I crawled up close to the base, with the bamboo rising and bowing over my head, looking down where I had last seen it. Then, glancing up, there it was right over my head, entwined in the bowed bamboo.

"Fuck!"

I jumped out. Again it disappeared. We lit a chunk of C4 plastic explosive and threw it into the bamboo, hoping to burn the snake out, but we didn't see it again. Soon we tired of the hunt and got back to the business of planning the following day's activities.

Consciously I put the snake out of my mind but it remained in my sub-conscious. That night I dreamed about it. In my dream, it had a head as big as a fist and it was really angry at me.

The next day dawned sunny and hot. One of the disadvantages of being next to a valley was that we were accessible to the general public. Around mid-morning, an ARVN (South Vietnamese soldier) came riding over to where we were on a Honda 50 motorcycle. He stopped and parked, using his M16 as a kickstand, muzzle down in the mud. Such lack of care for a weapon disgusted us and served as proof positive that the ARVN's were no good. One of my men came and got me. "Hey LT, Marvin the ARVN just showed up on his Honda 50. You wouldn't believe what he did. He propped his bike up with his M-16, muzzle down in the mud."

"Well, it's no wonder they're losing the fucking war. Where is he?"

He didn't seem to have any official reason for visiting us, or any reason at all for that matter. His manner was sort of surly and he reminded me of one of those slicky boys who rode around Saigon ripping people off. It also occurred to me that it would be quite simple for a VC to dress up in a set of fatigues and pass as an ARVN just to check us out. I was reminded why I preferred being in the mountains where there was no question about who was who.

Later in the day, a gang of kids showed up out in the valley about 100 meters from our position. It was clear that they wanted to come in and visit with us. I was equally clear that I didn't want them to visit us, for the simple reason that they could be working for the VC, reporting on our location, strength, positions, etc.

"Didi mau, didi mau!!" we yelled, waving them away.

But they wouldn't go away.

"Bugge, Brown, fire a couple of gas grenades out there, between us and them."

"Roger that, LT."

'BLOOP.'

The first grenade landed about twenty-five meters in front of them, letting loose a cloud of tear gas which, due to the absence of any wind, simply hung in the hot, humid air.

With shrieks of laughter, the kids ran away from the gentle drift of the cloud.

"Put another one out there, a little closer this time."

'BLOOP.'

Again, the kids laughed as they engaged in what for them was a great game of dodge ball.

I wasn't willing to harm them and it was clear that they were not impressed with the tear gas, so, in exasperation, I gave in.

"Fuck!! OK, wave them in."

More shrieks of laughter as they came running across the field to our position. There were about ten of them, ranging in age from twelve or thirteen down to what looked to be five or six. There were two girls, maybe ten or eleven years old. They wore pajamas and conical hats. Very shy, they remained squatted down on the periphery. The oldest boys were wearing shorts and pith helmets, the oldest smoking a big cigar. The little boys wore pajamas and little pork pie hats. They were very cute. A couple of the kids had pretty nasty cuts.

"Doc! Come over here with your aid bag and see what you can do for these kids."

Doc Wierzalis came over and, with McCartney as his assistant, cleaned and dressed their various wounds. The older boys and the girls remained impassively on the sidelines while the little boys and soldiers gathered around in rapt attention. Eventually, the novelty of visiting us wore off and the kids pushed on to greener pastures.

To add to the excitement, a Vietnamese voice sounded on my internal platoon radio frequency.

"LT, there's a dink on the horn."

Not knowing what this meant but sensing exciting possibilities, I grabbed the handset. "Who is this?"

"I am VC. I want to Chieu Hoi (surrender)."

Since this was the internal platoon radio net, all four squad radio operators were listening in. While I was filled with excitement at the prospect of bagging a prisoner, the soldiers listening in were equally excited but in a different way.

"Yeh, we'll Chieu Hoi you, mother fucker."

"Shut up, let me talk to this guy. What is your location?"

"I am in Que Son Valley."

"Come on in, we'll fix your ass."

"Shut up, clear the net."

And so it went for another minute, until our erstwhile Chieu Hoi finally cried out in a frustrated voice, "GI, why you fuck with me."

And he was gone.

After studying the map of the valley we were adjacent to, I decided to set up an ambush on a trail out in the valley that seemed a likely route for the VC and NVA to travel on. After dark, I led a squad size patrol out into the valley to the spot I had picked. We set up and waited. After a couple of hours, a helicopter gunship appeared in the sky about five hundred meters from us. It appeared to be patrolling the valley from the air. All of a sudden, it started firing its mini-guns at a target on the ground below. The guns were firing so fast that the tracer rounds created a solid line, as if a garden hose was squirting fire.

I wasn't sure what type of detection equipment they had to find targets in the dark, but it occurred to me that whatever equipment they had, we would look no different to them than would the VC/NVA, and I didn't like the idea of being hosed down by their mini-guns.

"Let's get out of here," I said.

We packed up and made our way back to the platoon perimeter, calling ahead to make sure they knew we were coming.

It had been an eventful few days.

Then, a couple of days later, we were abruptly extracted from the field.

Vietnamese kids

SP4 Dennis Thorkelson

SP4 "Doc" Wierzalis

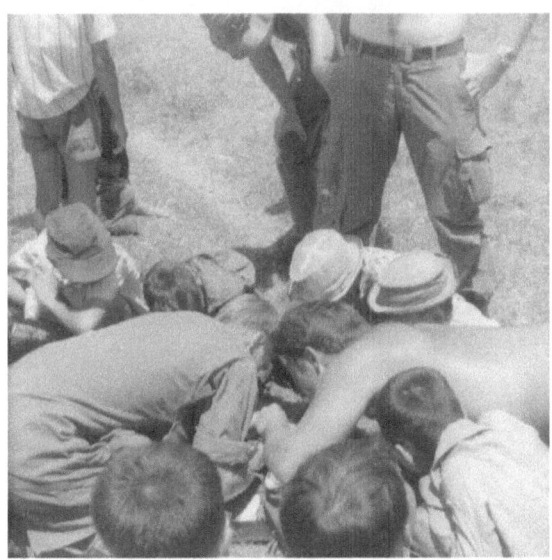

"Doc" Wierzalis and McCartney treating injured Vietnamese boy

Shy Vietnamese girls

Ni, our Kit Carson Scout

Older Vietnamese boy smoking a big cigar

Hai, our hooch maid

L-R SP4 Peter Gadzinski, SP4 Roberto Granado, SGT Rex Merrick, SGT Christiansen, and author

46. GOOD NIGHT AND GOOD LUCK

From the time I arrived in-country, the rumors had been flying about when we would be standing down. One rumor I remember was that a soldier was given a lift by the Brigade Commander, who confided in the soldier that we would be standing down on such and such a date. No matter how ridiculous and unlikely, such rumors would take off like wildfire. However, by early June, there were real indications that stand-down was imminent. The first tangible hint came in the first week of June when we returned to the battalion base camp from the mission on Charlie Ridge, only to find that the platoon leader's hooch was completely empty except for the three lockers belonging to the Charlie Company platoon leaders, which we had to empty and pack what we had in our duffle bags. I spent that night sleeping on the floor of the clerk's office.

The next day we went back to the field for what turned out to be our final mission. We were to patrol around the base of Firebase Linda while it was being dismantled. This was a necessary precaution but it was a pretty lame mission and no one had much heart for it, given that we were going to be pulled out at any moment.

That moment came rather suddenly when I received a call on the radio to prepare for immediate extraction. The soldiers were jubilant. None of them were going to be the last soldier killed in Vietnam. Once the birds had lifted off the Pickup Zone, everyone who had one popped a smoke grenade and held it out the open door of the chopper, so that each was spewing trails of red, yellow, and purple smoke as we made our way back to the base.

The base camp was a beehive of activity, which we were quickly sucked into. After turning in our weapons, we were instructed to deposit all our equipment in designated piles—helmets in one pile, canteens in another and so on. I gave all my food, G.I. and that sent from home, to our hooch maid, Hai. She was very sad at our leaving and I felt for her. As the widow of an ARVN soldier she had thrown her lot with us in order to survive and now we were abandoning her to whatever fate would befall those who had helped us in any way.

"I come see you ti-ti time," she said. It sounded like wishful thinking but who knows, maybe she made it to the States.

Once all our equipment was turned in, there wasn't much to do but wait our turn to do whatever was coming next. Some were going to the 3d Battalion, 21st Infantry, which would be remaining in-country a little longer. The rest would be transferred elsewhere in Vietnam or going home. All of us, enlisted and officers alike, were bedded down in a single hooch, sleeping on the floor. Orders for movement came unexpectedly, around the clock. A runner would come to the hooch and notify those whose time had come and they would just disappear. I remember Bugge waking me up in the middle of the night to say goodbye, and then he was gone. Gradually, my platoon, my family for the past six months, just drifted away like sand sifting through fingers. It was anti-climactic and, in a way, sad.

I only have snapshot memories of those few days.

- Sergeant Encarnacion insisting on spit-shining my paratrooper boots so that I would look sharp on the way home.
- A brief stand-down ceremony where the colors of our

battalion were furled, to be taken to Germany where the battalion would become part of the 1st Armored Division.
- A summons to brigade headquarters to see General McDonough in what turned out to be a social call. The connection between he and my father had somehow been established and he wanted to inquire after his old friend. On my return to base camp there was much joking among the lieutenants about me probably being the only lieutenant in the brigade to have his rights read to him by the brigade commander, followed by a social call.
- Paperwork to be completed and medals unceremoniously awarded.
- A quick trip into Da Nang to register a Chinese SKS carbine as a war trophy. Beautiful, airy French colonial buildings — a real contrast to the plywood hooches of our base camp, not to mention the poncho hooches we slept under in the field.
- While lying on the floor one night trying to go to sleep, hearing the crystal clear voice of Joan Baez wafting through the night air from someone's boom box and being positively enchanted.

Notification came that I would be on the next flight out, a daylight run to Saigon on 20 June. To my delight, my good friend John Beasley was on the same manifest, which meant that we would be making the trip home together. Over fifty of us were piled on trucks and transported from Camp Crecenz to Da Nang Air Base. When it came time to load on the plane, a C-130 cargo plane, the fastidious and procedure-conscious Air Force did something I never experienced before or since — they simply packed us into the cargo bay of the aircraft without benefit of seats, safety restraints, or anything. We just got in as tight as we could and sat on the floor. None of us cared but it was most unlike the way the Air Force normally operated. Once aboard, the rear ramp was closed and we took off, headed for Saigon where we would catch a Freedom Bird at Ton San Nhut Air Base.

Once the C-130 was airborne, there was collective joy that we were actually on our way home.

After flying uneventfully for about an hour, the plane banked sharply—abnormally so, and a shock of concern brought all of us to full alert. The pilot came on the intercom,

"We've encountered a problem and we need to return to Da Nang. Sorry 'bout that."

Euphoria quickly gave way to dismay and dread of what we were returning to. Was the plane in danger of crashing? Did the NVA or VC attack? None of the possibilities sounded good. After flying for another hour, we landed, taxied, and then came to a halt. We sat in the semi-darkness of the cargo hold waiting for the rear ramp to be lowered and wondering what had gone wrong and what that meant for us. Finally, the ramp lowered and we were hit with the bright sun of a Vietnamese hot-season day. We stumbled off the plane into the blinding sunlight, looking around while adjusting our eyes to the light. All we could see in any direction was airfield, undistinguishable from any airfield anywhere. As we waited, uncertain as to our fate, a bus arrived and we were welcomed to …. Saigon. The pilot had played a major joke on us—and a good one it was with its happy outcome.

We were taken to Camp Alpha, which was a transit camp similar in function to the one I stayed at upon arrival in country. Unlike the typical wood hut construction, the buildings in this camp were made of concrete, two-storied and arranged around a central courtyard. They were shells only—no doors, window-panes or finish work of any type. The only amenities were toilets, showers, and bunks in the sleeping bays, which was fine. Flush toilets were a luxury we had not known up-country.

That night John and I went to a show at the Officer's Club. The group giving the performance seemed to be able to sing in all genres, from pop to country and western. The guys from the 196th were sitting on the right side of the theater, and over to the left were some guys from the 1st Cavalry Division, with their distinctive yellow horse-blanket shoulder patch with the black horse head on it. After several beers, we began hurling insults at them.

"If you ain't a grunt, you ain't shit."

"Fuck the Cav. The line you never crossed, the horse you never rode and the color speaks for itself."

The Cav men were yelling back at us but things never progressed beyond insults and name-calling. Later I sent a request up to the stage to sing the Merle Haggard song, *"I take a lot of pride in what I am"* and to dedicate it to the grunts of the 196th Light Infantry Brigade.

> *"Things I learned in a hobo jungle were things*
> *they never taught me in the classroom…"*
> *"Never been nobody's idol, but at least I got a title*
> *And I take a lot of pride in what I am*
> *And I take a lot of pride in what I am."*

Wild cheers.

"Grunts to the max!!"

"If you ain't a grunt, you ain't shit!"

***

Time to board the Freedom Bird. Dressed in khakis with medal ribbons and Combat Infantryman's Badge proudly affixed above left breast pocket, we filed quietly to the chartered airliner which would take us back to "The World". Once aboard and seated, there was a palpable feeling of suspense. There was little talking and that which did occur was done in the hushed tones reserved for church or funeral parlors. The jet engines revved up and the plane started to taxi down the runway. Quieter still. Everyone seemingly holding their breath. Front wheel off the ground, plane tilted upward. Back wheels leave the ground. We are airborne and with that erupts a mighty exhale and cheer.

The flight home was uneventful. Boots off to allow for swelling feet. Lots of visiting up and down the aisle with Sergeant Rodriguez, Gadzinski, and other comrades. The initial euphoria gave way to fatigue and most slept. Amongst the many useful skills learned in military training is the ability to sleep anywhere, any time.

At about 2:30 a.m. we received the alert that we would soon be landing at Travis AFB. We quickly tidied up our personal appearance as best as we could after having been in the air for eighteen hours. After landing and taxiing to the terminal, the doors were opened and we filed out into...freezing weather! I had forgotten and most of the others didn't know how cold California nights can be, even in the summer. As soon as we were re-united with our kit bags, there was a mad digging to find the field jackets which had seldom if ever been worn in Vietnam.

At this point, there was a further fragmentation of our group of returnees. Some boarded buses which would take them straight to the San Francisco airport. Others boarded buses destined for Oakland Army Terminal. Since my parents lived in the Bay Area, John Beasley and I caught the bus for Oakland where we arrived at 4:00 a.m. I didn't want to call my parents at that ungodly hour, so John and I partook of the steak dinner which was served 24 hours a day to Vietnam returnees. After that we kicked around until 6:00 a.m., which I deemed was a more reasonable time to awaken my parents. They were over the moon happy that I was safely home and my dad was there within 30 minutes to pick us up. Back at the house, dad cooked up some breakfast, no doubt his signature scrambled eggs made with whole cream and thick sliced bacon. We sat around the living room making small talk and then my dad disappeared upstairs. When he came back down he walked over to where I was sitting and said, "Here, I guess you should have this."

It was his Combat Infantryman's Badge from the Korean War. Words fail to describe the significance of this exchange. Maybe only an infantryman would understand.

John was anxious to get home to Chicago so we piled back into the car and drove to the San Francisco airport. These were pre-security days and you could accompany passengers right up to the boarding gate, which I did. When it came time for John to board his flight, I stepped up to say goodbye to the most consistent friend and comrade of my short military career. I extended my hand to him in the manner peculiar to soldiers in Vietnam. He extended his hand in the way that is "normal". When he saw that I had extended my hand in the "Nam" way, he switched just as I switched to the "normal" way. Things were changing. Physically, we had left one world and were in another. In other ways, part of us was in one place and part of us was in the other.

APPENDIX A

CAST OF CHARACTERS

When I think of the platoon I think of a fixed, unchanging group of guys who are forever imprinted in my memory as "The Platoon". The reality is that a lot of people came and went. The bulk of the platoon that I remember consists of those who were there when I took it over, and those replacements who were there up to the end. Following are the members of "The Platoon". My apologies to any whom I have forgotten:

SGT Russell Christianson, Tacoma, Washington 1st Squad Leader
SGT "Reincarnation" Encarnacion, Puerto Rico
SSG David Gauthier, Maine Platoon Sergeant
SGT David "Rock" Mixon, Texas
SGT Ricardo "Rod" Rodriguez," Laredo, Texas Platoon Sergeant/ 2d Squad Leader
SGT "Stringbean" Stringfellow, Ft. Worth, Texas
SGT Stephen Willard, Detroit, Michigan
PFC Bubb Anderson, Bastrop, Louisiana
SP4 Danny Bishop, Arkansas
SP4 Bourland
SP4 Jesse "Blooper" Brown, Swainsboro, Georgia

SP4 Ruben Bugge, San Rafael, California
SP4 Thomas Carlson, RTO, Denver, Colorado **SP4 "Doc" Cousino,** Ohio
SP4 Cecil Dodd, Spartanburg, South Carolina
SP4 John Elzey, Clinton, Mississippi
SP4 Peter Gadzinsky, RTO, Chicago, Illinois **PFC Gallapo**
PFC Richard Garcia, Los Angeles
SP4 Roberto Granado, Glendale, Arizona
SP4 Ray Hanisco, Pig gunner, Pennsylvania
PFC Houchin, Tunnel Rat, Oklahoma **Hartwig Jackson**
PFC Victor Jeffries, Indiana
SP4 "Doc" Kroze, Everett, Washington
PFC Robert McCartney, Demolitions man, Cheyenne, Wyoming
SP4 James F. "OB" O'Brien, Springfield, Massachusetts
SP4 Steven Olsen, Iowa
SP4 Danny Salazar, Palm Springs, California
PVT "Smitty"Smith, Point man
SP4 Melvin Smoots, Vicksburg, Mississippi
PFC Eddie Sosa, Dallas, Texas
SP4 Tetrault, New Hampshire
SP4 Dennis Thorkelson
PVT Charles "War" Warner , Colorado
PFC "Doc" Wierzalis, Philadelphia, Pennsylvania

Although the vast majority of the guys in the platoon were good men with whom I was proud to serve, I would be hard pressed to say something interesting about each one of them. Following is a sampling of the cast of characters, to include some outside the platoon who were, nonetheless, part of the picture:

Lt Col Clyde Tate. Lt Col Tate was the battalion commander when I first joined the battalion. I don't know if he actually was good but he walked and talked the part and seemed like a good CO. He had joined the Army at seventeen and worked his way up through the ranks and was on his second combat tour in Vietnam. When you are young, appearances count for a lot. Part of Colonel Tate's image was that he carried a Swedish K sub-machine gun. One might well ask why a battalion commander needed to carry any weapon, but to my young eyes this gave him an aggressive, swashbuckling mystique.

Early in my tour, Charlie Company was air-assaulted onto Da Nang Air Base as a practice maneuver for defending it in the event that it came under direct attack (which had happened in the past). Afterwards, Col Tate gathered the company together and told us that we were his best company and that he knew he could count on us when the chips were down. Of course, this made us feel very proud. Later, talking with some of my friends who were lieutenants in other companies, I found out that he told all the companies the same thing. It was a good ploy, though, because the various companies in the battalion seldom saw each other.

Lt Col Andrew D. Perkins. Lt Col Perkins was Lt Col Tate's successor to command of the battalion.

Maj. George "Bruiser" Lenhart. The "Bruiser" was the battalion Executive Officer or second in command to the battalion commander. As such, he was in charge of the base camp and all administration and logistics. His nickname came from his passion for athletics, his physical appearance and demeanor. He was barrel-shaped with great, hairy bowling pin arms. He was given to inspecting our hooch when we were in from the field and haranguing us for our lack of neatness. In one of life's ironies, George "Denny" Lenhart lives five miles from me and we have become good friends. I like to think of this is as God's way of letting me 'get even'.

Captain William Thomas, Commanding Officer, Co. C. Capt. Thomas was a former NCO who had gone to OCS and been commissioned as an officer.

Captain "Crunch". I don't even remember his real name. He arrived in Vietnam at the same time as I did and was fresh in the Army from ROTC. There were not enough lieutenants to command the platoons in the company so he was given to Sergeant First Class Rollins, the most seasoned platoon sergeant in the company.

Sergeant First Class Randall Rollins, "The Bear". SFC Rollins was a big hulk of a man from West Virginia, He was a Korean War veteran and had been with the 1st Cavalry Division during some of its heaviest fighting earlier in the war. He was gruff but likable — to me, anyway. I think the troops liked him as well in their own way. They certainly respected him. Profanity gushed from his mouth like water from a spring, so much so that it was funny. Once when we were in the battalion base camp and SFC Rollins was holding a company formation, I saw a soldier in the rear rank tape recording him — not out of malicious intent but because what came out of his mouth was so colorful and amazing. He was happiest in the field and could hump the mountains like he was walking down the street.

Sergeant Russell Christianson. Sergeant Christianson was a hard-stripe buck sergeant who led 1st Squad. He was a very competent and reliable NCO who had been in Vietnam for a while. His only drawback was that he had a very volatile temper and would have the occasional temper tantrum when things didn't go his way. That was the main reason that I didn't choose him to be the platoon sergeant when there was no senior NCO for that position. Lest I create the wrong impression, he also had a good sense of humor and was a mainstay of the platoon.

He and Rod used to snipe at each other a lot over hard stripe NCOs versus 'shake and bake' NCOs.

Sergeant Ricardo Rodriguez. Rod was normally the 2d Squad Leader but for much of my tour he was the platoon sergeant. The platoon sergeant, or "Plat Daddy" is normally a Sergeant First Class with 12—15 years of service. Rod had been in the service about as long as me, maybe less but he had a natural competence that made up for his lack of experience. He was a product of the NCO Candidate Course at Ft. Benning. Recruits with high scores were given the opportunity to attend this course and graduate as buck-sergeants. The course was started to make up for the shortage of NCOs in Vietnam. Inevitably there was tension between sergeants like Christianson who had worked their way up the ranks and 'Shake 'n Bake' NCOs like Rod. He could get the job done without being a hard-ass about it, which was an important attribute at that time in the war. He also had a playful sense of humor. As I mentioned earlier, one time when we were manning the Ridgeline, I received a package of dried fruits from home. My favorites were the dried apricots. One day I returned to the bunker we shared and got out the bag of fruit, only to find that all the apricots were gone. Rod had eaten them all. I was mad and he laughed. Vietnam was funny like that. On the one hand, what we were doing was deadly serious and soldiers were ordered to do very dangerous things. On the other hand, we were like a big family where something like this prank could happen without lessening the fundamental discipline of the unit.

Sergeant "Rock" Mixon. Sergeant Mixon came to me from the brigade Ranger company and was airborne and Ranger qualified. I couldn't believe my luck to have gotten such a jewel. But looks can be deceiving and all that glitters is not gold. For all his qualifications, SGT Mixon was lazy, dishonest, and unreliable. He was certainly capable or he wouldn't have made it through Ranger School. When given a patrol, he wasn't above hiding somewhere and calling in fake situation reports on the radio. He was also cruel and I knew that he would jump at the chance to kill someone. I once warned him that I had my eye on him and that he had better not even consider killing someone in cold blood.

SP4 Jesse "Blooper" Brown. Brown got the nickname "Blooper" because he was a grenadier. The grenade launcher was also called a 'Blooper" because of the sound it made when it was fired. Brown was a quiet unassuming soldier but was thoroughly reliable and trustworthy. I submitted his name for promotion to sergeant but, unfortunately, we were so flooded with "Shake 'n Bakes' that he never got the promotion. He would have made the kind of quietly competent NCO that we needed. He was a good man.

SP4 Ruben Bugge. Bugge was sort of a carefree hippy from San Raphael, California, but a good, reliable member of the platoon. When we were in the rear, he and Warner were given to making late night visits just to chew the fat, have a smoke and maybe a beer.

PFC Cecil "Doc" Dodd. Dodd was what we called our OJT (On the job training) medic. The medics had trained him to do many of the routine medical treatments and he assisted the platoon medic as necessary. His basic job was that of rifleman but he carried a full medic's aid bag in addition to all his other gear. He was a very nice man who was always ready with a huge smile.

SP4 John Elzey. Elzey was another one with leadership potential who would have made a good squad leader had his promotion gone through. He was an acting squad leader on occasion when there were no sergeants available to do the job. Even though I often had plenty of 'Shake 'n Bakes' around, some of them were not to be trusted with a squad and I would use a good SP4 before using them. Elzey and Brown were two such men.

SP4 Peter Gadzinski. Gadzinski was my RTO (Radio Telephone Operator). Looking back, I realize that there is a difference between comrades and friends. Comrades are people with whom you share a special bond because of what you have experienced together. Beyond that bond, you may have nothing in common. Gadzinski, I would have to say, was also a friend, even though military protocol wouldn't allow either of us to think of our relationship in those terms. When time allowed, we would have long conversations at an intellectual level that would have been impossible with most of the men. I contacted him in 2012, thanks to computer technology and the uniqueness of his name.

PFC Garcia. Garcia was one of the quartet of Mexicans who were in the platoon when I arrived. He was a good soldier but not as good as the other three—which is to say that the Mexicans tended to make very good soldiers. What was unique about Garcia was a certain fastidiousness about his appearance, which was difficult given the nature of our lives. His penchant to be a Beau Brummel of the Bush was most exemplified by his insistence on getting a new camouflage cover for his helmet every time we returned to the base camp. Everyone's camouflage covers were faded and ragged and there was a certain pride about that among most grunts, but not Garcia.

SP4 Roberto Granado. Granado was twenty-two years old, was married and had kids. This made him an old man among boys. He was the moral leader of the Mexicans, including Rodriguez. What leadership he exercised was quiet, unobtrusive, and effective. Based on the recommendations of several members of the platoon, I recommended him for a Bronze Star for valor. I am in touch with him to this day.

PFC Houchin. Houchin was a country boy from Oklahoma, very rustic. He was also the platoon tunnel rat. There was no hole he would not crawl into, even if it was only wide enough for his elongated body to wiggle into. It gave me claustrophobia just to watch. His favorite catch phrase was, referring to the M16 rifle, "You take care of it and it will take care of you." He would share this nugget of wisdom as an old man would with a youth.

SP4 "Doc" Kroze. Doc Kroze was one of three medics that served in the platoon. He was interesting because he was a conscientious objector. I don't remember what his religion was but whatever it was he refused to carry a weapon. He was a good medic and pulled his weight in the platoon and his decision to not carry a weapon was respected.

PFC Robert McCartney. McCartney was in the company when it was over-run by the Viet Cong on Firebase Mary Ann some months earlier. Following that he went AWOL and had now returned to finish his tour. Unlike the stereotype of a disgruntled soldier gone AWOL, McCartney was cheerful. He also liked to blow things up and was the platoon demolitions man. He was short and stocky and could throw a white phosphorous grenade farther than anyone. (WP grenades were quite heavy).

PFC James F. "OB" O'Brien. O'Brien was the quintessential Irish American kid that appears in every Hollywood war movie. Broad Irish face covered with freckles and topped with a thick bush of reddish blond hair. When he arrived in the platoon as a replacement, he was an innocent, fresh-faced kid who was eager to please and do his best.

As time went on, I noted a hardness developing in him, which in a way was sad to see. Innocence lost.

PVT Charles "War" Warner. Warner was a wild, happy go lucky cowboy type from Colorado. He had apparently been in Vietnam for years and had been a Staff Sergeant at one time. I never learned or don't remember all the particulars of his bust down to Private, but he was a good soldier who frequently volunteered to walk point. He seemed to like what he was doing, as indicated by a tattoo that said "War". He also wore on a wide leather wrist strap a medallion which said "War". When we were in the rear, he and Bugge would sometimes come visit me in the middle of the night to have a smoke and shoot the bull.

Ni. Ni was the Kit Carson Scout assigned to my platoon. A Kit Carson Scout (or Kansas City Star, as we sometimes phonetically referred to them) were former Viet Cong or North Vietnamese Army soldiers who had switched sides, or come over to us under the Chieu Hoi program. In my limited experience, KCS's who were formerly NVA were tougher and more reliable than those who were former VC. Ni was quite young and had been an NVA soldier. As he didn't speak English nor I Vietnamese, I can't say too much about him except that he was very friendly and willing.

Hai. I would be remiss to not mention Hai. She was the hooch maid for the lieutenant's hooch. My understanding was that she was the widow of an ARVN soldier who had been killed in the war. The deal was that the hooch maid would keep the hooch clean and would take care of each lieutenant's clothing and boots in exchange for a small fee. I balked at this at first out of my egalitarian attitudes about not having someone else do for me that which I could do for myself. I also balked at parting with my brass, as I was only receiving $35.00 per month, the rest going home in an allotment. However, whenever I came in from the field, my fatigues were lying on the bunk neatly pressed and my other pair of boots was cleaned and shined. I couldn't stop Hai from doing these kindnesses and my sense of fairness dictated that I compensate her. So I ended up paying her pretty well for her work. She was very sad when we left and I felt sorry for her because she had thrown her lot with the Americans and we abandoned her to her fate at the hands of the victorious North Vietnamese and Viet Cong.

Having written this summary of the men in the platoon, I am struck with a couple of thoughts. One is that they were a good cross-section of Americana, both geographically and racially. Secondly, they were anything but the wild-eyed, dope crazed mutinous mob commonly depicted in press and film. In my memory, they were just a bunch of guys doing a job they didn't want to do but doing it to the best of their ability. And for that they have my undying respect.

APPENDIX B

WHAT WE CARRIED

We were light infantry, which, from a strategic perspective, meant that we could operate in any terrain, unencumbered by a logistical tail of trucks or armored personnel carriers which would transport us and carry our equipment. We were particularly suited for penetrating mountainous terrain, which was inaccessible to wheeled or tracked vehicles. This meant that everything we needed to survive and fight with was carried on our backs. Paradoxically, the "Light Infantry" carried very heavy loads. This is what we carried.

From the inside out: Two dog tags on a chain, an OG (olive green) cotton undershirt, and OG cushion sole wool/nylon boot socks. No drawers. No one wore drawers because they would get soaked with sweat and cause chafing of the crotch and inner thighs, which was more debilitating than blisters on the feet. It was a moot point anyhow, because the drawers disappeared the first time they were sent to the laundry. Presumably, the Vietnamese coveted them as shorts.

The next layer consisted of a steel helmet (2.85 lbs.), jungle fatigues, a roomy, lightweight field uniform which featured four big pockets on the jacket and two large cargo pockets on the trousers, and jungle boots, into which were laced a third dog tag. On separate lanyards around my neck were a lensatic compass and the code-book (SOI) which contained all the radio frequencies and call signs, as well as the codes for encoding and de-coding messages.[13]

In the right cargo pocket of my trousers was an extra 30-round magazine for my rifle (1 lb.) and in the left pocket was my map and a book, both in waterproof plastic bags.

Over the uniform went LBE (Load Bearing Equipment), which consisted of a thick, webbed belt to which was attached heavy canvas suspenders (5 lbs.). To this was attached two ammunition pouches, each holding four 20 round magazines (16 lbs.), two canteen pouches with canteen and cup (4 lbs.), first aid pouch, two hand grenades (2 lbs.), two smoke grenades (2 lbs.), and one thermite grenade (1 lb.). Weight: 30 lbs.

Next came an OG towel, which was draped over the shoulders for use in wiping away sweat and providing some padding against the straps of the rucksack. Into the rucksack went the gas mask, monsoon sweater, poncho and poncho liner, 2–3 extra pairs of socks, rations for three days (7.5 lbs.), two bags or bottles of plasma (4 lbs.), and the demolition kit, a wooden box containing electric and non-electric blasting caps and a detonator. Attached to the outside of the ruck were two two-quart canteens (8 lbs.), entrenching tool, a bandolier of seven 20 round magazines (16 lbs.), a LAW (Light Anti-tank Weapon (5 lbs.), and a machete. To the bottom of the ruck was attached a water tight ammunition can, called a 'ditty box', in which was stored mail, writing paper, etc. Rucksack weight: approximately 50 lbs.

---

13. The SOI (Signal Operating Instructions) was a very sensitive item in that it contained the radio frequencies and call signs for the entire brigade for a whole month. The frequencies and call signs changed every 24 hours, usually at midnight. If the NVA had access to this information, they would be able to discern our every move. The loss of an SOI would necessitate the entire brigade changing SOI's all the way down to platoon level. For that reason, everyone who had an SOI had it secured to their person by a nylon cord.

Carried in hand was one's individual weapon, in my case an M-16 rifle (8.5 lbs.).

Total Weight: Approximately 90 lbs.

Every soldier carried some variation of this, some much heavier. Radio operators carried a PRC 77 radio, (14 lbs.), plus an extra battery (3 lbs.). Machine gunners carried the M-60 Machine Gun (23 lbs.) and a belt of 100 rounds of ammunition (7 lbs.). Medics carried on top of their ruck a complete medic's aid bag (21 lbs.). Others carried blocks of C-4 explosive, claymore mines, and additional ammunition for the machine guns.

It was pretty grueling, but we were young.

APPENDIX C

HOW WE OPERATED

Before beginning my description of how we operated, it is important to point out that this is a description of how we operated at this location at this phase of the war. At other times and in other places during the war, units operated quite differently.

The brigade Area of Operations (AO) was divided into two battalion-sized AO's, named AO Maude and AO Linda. Each of these AO's included a segment of the Annamite Mountains as well as the foot-hills and valleys between the mountains and the Da Nang Air Base. Centrally located within each was a hilltop firebase which contained the Battalion Tactical Operations Center (TOC), the battalion mortar section and a five-howitzer artillery battery. From the firebase, the battalion commander could provide command and control of his infantry companies as well as indirect fire support throughout the AO. The battalion commander would further sub-divide the battalion AO so that each of the four line companies (A, B, C, and D) had its own area to work. Company commanders would again sub-divide its area so that each of his three platoons had a distinct area in which to operate. Battalions were assigned to work these AO's for a period of 44 days, at the conclusion of which they would be rotated to the rear for other missions, or transferred to the adjacent AO.

At the beginning of a mission, the company would fly from our base camp into our assigned AO. If we were starting out at the lower elevations, which consisted of grassy foothills and valleys, the landing zones (LZ) would accommodate five helicopters ("slicks") at a time. It took five helicopters slicks to lift a platoon so it would take three trips to get the whole company to the field. If we were beginning the operation in the densely-wooded mountains, the LZ would probably be a one-shipper and it would be a lengthy process to get the whole company on the ground. Once on the ground, each platoon would take off in a different direction and that would be the last that we saw of each other until the mission was over. The company commander would accompany a different platoon on each mission. During that forty-four day cycle we would be brought to the rear every few weeks to have a shower, get a clean uniform, and go back to the field. We would come in one day and go back out the next. In many ways, it was more trouble than it was worth to come in at all but that's the way it was.

Our main objective was to prevent the NVA from rocketing the air base and, secondarily, to interdict the flow of enemy soldiers and materiel from the mountains to the populated areas on the coast. The Ho Chi Minh Trail could be likened to an enormous river down which flowed men and equipment from North Vietnam along the border with Laos and Cambodia into the south. From this main "river" channel there were dozens of branches, like tributaries in reverse, which led to the various population centers of South Vietnam. The major branches were what we referred to as "super-highways"—wide, well-traveled, meticulously camouflaged routes which could even accommodate wheeled vehicles. From these branches were numerous tiny trails which filtered out of the mountains onto the valley below.

Once we were on the move as platoons, we would usually move six to eight kilometers until early afternoon. Route selection was very important. Traveling by stream or existing trail was the easiest and most likely to result in evidence of enemy activity, but it was also dangerous. Cutting trail through the jungle was the safest but was exhausting and was not particularly helpful in finding the enemy. All these considerations had to be taken into account when choosing a route.

In early afternoon, I would select a defensible spot for a patrol base. We would set up a perimeter and drop rucksacks. From here each squad would be sent out on a looping patrol route which would bring them eventually back to the patrol base. With each squad performing this "cloverleaf" pattern, the entire area surrounding the patrol base could be covered. I would usually go with the squad most likely to make contact.

Once the patrols were completed, we would saddle up and move until early evening when I would select another defensible spot, usually a knoll or hilltop, where we would set up a Night Defensive Position (NDP). Each squad would be assigned a sector of the perimeter. I would direct the positioning of the machine guns on the most likely avenues of approach as well as identifying dead space to be covered by the grenadiers. At least three ambushes would be laid; two Mechanical Ambushes (MA's), which were activated by a trip-wire and one manned ambush, usually on a nearby trail or stream. At night under normal circumstances, thirty per cent alert was maintained, going up to fifty or one hundred percent if the situation dictated it. When I slept, I stayed fully clothed with my boots on but with the laces loosened to ease circulation. My rifle would be cradled in my arms. We usually went to "bed" soaking wet either from rain or sweat. It was easiest to sleep in the monsoon season when it could get quite cool at night and there weren't as many mosquitoes. One could wrap up in ones camouflage nylon blanket and be reasonably comfortable. In the hot season, sleep was virtually impossible with the heat and clouds of mosquitoes being a constant torment.

In the morning, the platoon was at one hundred per cent stand-to thirty minutes before dawn. At dawn the MA's were brought in and the manned ambush would come in. We would eat breakfast, which, in my case, was a can of C-ration fruit and a cup of coffee drunk out of a C-ration tin, and then we would saddle up and move out. With variations, this process was repeated every day.

Unlike the old adage that war consists of prolonged periods of boredom punctuated by moments of intense fear (and exhilaration), I would characterize our life as being prolonged periods of grinding fatigue accompanied by high levels of alertness and low levels of fear, punctuated by moments of intense fear and exhilaration.

ABOUT THE AUTHOR

Byrne N. "Buzz" Sherwood was commissioned as a lieutenant of infantry upon graduation from LSU in 1970. In 1971, he participated in the last phase of the war in Vietnam as a rifle platoon leader in the 196th Light Infantry Brigade. Serving for twenty-three years, the capstone of his career was command of the 2d Battalion, 22d Infantry, 10th Mountain Division. Following retirement from active duty, he taught high school in Richmond, California. He is currently a volunteer chaplain at the Alameda County Juvenile Detention Center.